REACHING NET ZERO

REACHING NET ZERO
What It Takes to Solve the Global Climate Crisis

WILLIAM D. FLETCHER
Rockwell International Corporation, (Retired) Costa Mesa, CA, United States

CRAIG B. SMITH
DMJM + H&N, (Retired) Los Angeles, CA, United States

Elsevier
Radarweg 29, PO Box 211, 1000 AE Amsterdam, Netherlands
The Boulevard, Langford Lane, Kidlington, Oxford OX5 1GB, United Kingdom
50 Hampshire Street, 5th Floor, Cambridge, MA 02139, United States

Copyright © 2020 Elsevier Inc. All rights reserved.

No part of this publication may be reproduced or transmitted in any form or by any means, electronic or mechanical, including photocopying, recording, or any information storage and retrieval system, without permission in writing from the publisher. Details on how to seek permission, further information about the Publisher's permissions policies and our arrangements with organizations such as the Copyright Clearance Center and the Copyright Licensing Agency, can be found at our website: www.elsevier.com/permissions.

This book and the individual contributions contained in it are protected under copyright by the Publisher (other than as may be noted herein).

Notices
Knowledge and best practice in this field are constantly changing. As new research and experience broaden our understanding, changes in research methods, professional practices, or medical treatment may become necessary.

Practitioners and researchers must always rely on their own experience and knowledge in evaluating and using any information, methods, compounds, or experiments described herein. In using such information or methods they should be mindful of their own safety and the safety of others, including parties for whom they have a professional responsibility.

To the fullest extent of the law, neither the Publisher nor the authors, contributors, or editors, assume any liability for any injury and/or damage to persons or property as a matter of products liability, negligence or otherwise, or from any use or operation of any methods, products, instructions, or ideas contained in the material herein.

British Library Cataloguing-in-Publication Data
A catalogue record for this book is available from the British Library

Library of Congress Cataloging-in-Publication Data
A catalog record for this book is available from the Library of Congress

ISBN: 978-0-12-823366-5

For Information on all Elsevier publications visit our website at https://www.elsevier.com/books-and-journals

Publisher: Candice Janco
Acquisitions Editor: Peter J. Llewellyn
Editorial Project Manager: Chiara Giglio
Production Project Manager: Kumar Anbazhagan
Cover Designer: Mark Rogers

Typeset by MPS Limited, Chennai, India

Contents

List of Figures	xi
List of Tables	xiii
Photographs	xv
Preface: Why read this book?	xvii
Acknowledgments	xxiii
Acronyms	xxv

1. Introduction — 1

Should we be concerned about global warming?	1
What about solar radiation?	2
The greenhouse effect	2
What are the greenhouse gases?	3
What are the signs of global warming?	5
What are the dangers of global warming?	5
Can anything be done about global warming?	7

2. Addressing global warming — 9

Latency is a huge problem	11
Global warming is not obvious to the vast majority of people	12
The global economy is powered by fossil fuels	12
There is a need for unprecedented and perhaps unachievable international cooperation	13
All of us will have to be willing to accept changes	14
U.S. participation is essential	14
What can be done?	14

Part I

3. The earth as a system — 21

Incident solar radiation	21
Milankovitch cycles	23
More about the greenhouse effect	25
Carbon cycle	26
Temperature increase	30

4. Fundamental drivers of global warming — 31

Global population rising	31

v

Inequities: the early role of the United States and the United Kingdom	31
The pivotal position of the United States	33
Need to consider both absolute and per capita emissions	34
Gross domestic product growth and energy use are related	36
More energy will be required by developing countries	37

5. How do we know global warming is real?   **39**

Global warming is not a new idea	39
CO_2 emissions are rising	40
Earth's temperature is rising	40
Correlation of increasing global temperature with increasing atmospheric CO_2	41
Ocean temperatures are rising	42
Sea levels are rising	43
Ocean acidification is occurring	45
A message from the Arctic?	45
Glaciers, ice caps, and sea ice are melting	47
The permafrost is melting	54
Extreme weather events are increasing	54
Deserts and tropics are expanding	55
Rising temperatures are causing plant, animal, and human migration	58
Early warning signs of global warming: a California case history	58

6. How do we know man-made CO_2 is the issue?   **61**

Where do man-made greenhouse gases come from?	62
What happens to CO_2 emissions?	65
The significance of carbon-14	68
Historic emissions since the Industrial Revolution	70
Increases in atmospheric CO_2 correlates with fossil fuel use	72
What is your carbon footprint?	73

7. What are the effects of global warming?   **75**

Latency—how long before effects show up?	76
Climate change versus weather	76
Earth's temperature will continue to rise	78
Air pollution will increase	79
Sea levels rise causing flooding	79
Oceans become more acidic	83
Glaciers, ice caps, and sea ice melt	84
Subsidence occurs and permafrost melts	85

Contents

vii

Deserts and tropics expand	87
Species migration and extinction	87
Frequency and severity of storms	89
Impact on agriculture, droughts, loss of cropland, and wildfires	90
Health problems will be more severe	92
Could global warming cause a financial crisis or some other financial problem?	94
National security implications	95
Migrations caused by climate change	96
Tipping points: unanticipated changes can occur	97

8. International efforts to address global warming
99

Early efforts	99
The Intergovernmental Panel on Climate Change	100
The Paris Agreement	101
Intergovernmental Panel on Climate Change special reports	102
History of Intergovernmental Panel on Climate Change global warming objectives	103

Part II

9. What would it take to reach net zero?
107

Intergovernmental Panel on Climate Change alternative scenarios	107
What would it take?	110
Are the Intergovernmental Panel on Climate Change scenarios realistic?	111
Carbon removal	115
What is a more likely scenario?	115
Are we too late already?	116
Doing nothing is not an option	117
What will happen if we do nothing?	118
The high cost of doing nothing	121

10. Energy alternatives
123

Fossil fuels: coal, oil, and natural gas	125
Nuclear power	128
Renewable energy	129

11. Unique problems of major contributors to global warming
149

What can we learn from Germany?	149
The United States fails to take a leadership position	153

viii Contents

China—Will it be the leader? 156
India—large population, little energy 158
Japan—strong technological capabilities 159
Russia—may not be a player 160
Observations 161

12. Why is global warming such a difficult problem to solve? 163

The need for unprecedented, perhaps unachievable, global cooperation 163
Fossil fuels are heavily subsidized 164
Educating the public 164
The media have not dealt fairly with global warming 165
Public uncertainty 167
A positive message is needed 169
Public support for government action 169
Why it is hard to replace fossil fuels? 170
Solving technical challenges 171
The need for strong economies 173
Understanding climate change skepticism 174
Recognizing political leaders can make mistakes 176
Acknowledging that failure is a possibility 176

13. Some successes and failures 179

The Permian Basin, a renewable energy powerhouse 179
1970s oil price hikes 181
Automobile emissions 184
Hole in the ozone layer 185
Cigarette smoking and cancer 186
Europe's push for diesel vehicles 187
Nuclear power in the United States 188
Is there a future for nuclear power? 191
Ethanol 195
High-speed rail ´97
Lessons learned 198

Part III

14. Action Plan: efficiency, power, transportation, and land use 201

Do we need another moon shot? 201
The challenges of a global approach 202

Contents   ix

An Action Plan, assuming we cannot get to net zero by 2050   203
Can the Intergovernmental Panel on Climate Change's goal of keeping global
warming under 2°C be met?   214
Why can't we do better?   216
Silver bullets   218
Mitigation   219
Carbon fee   223

15. Can it be done?   **227**

The trend is our friend   227
Can renewable energy power the world?   230
Can wind power the world?   232
What would it cost?   233
Can we afford it?   234

16. The way forward   **239**

The future can be bright   240
Top priorities   241
Government actions   242
Actions for concerned citizens   250
Actions for industry   253
What next?   257

Afterword   *259*
Further reading   *261*
Useful reports   *263*
Useful websites   *265*

Part IV Appendices

Appendix 1: Abbreviations, units, and conversion factors   *269*
Appendix 2: The amount of CO_2 in the atmosphere: sources and sinks   *275*
Appendix 3: Will the IPCC goal of 450 ppm be met?   *277*
Appendix 4: Key parameters used to formulate Action Plan   *279*
Appendix 5: Flood and sea rise mitigation   *285*
Appendix 6: Financial measures   *289*
Appendix 7: Activist and lobbying groups, litigation examples   *293*
Appendix 8: Excerpts from corporate annual reports   *301*

Index   *307*

List of Figures

Figure 2.1	An example of latency	13
Figure 3.1	Solar energy sources and sinks	22
Figure 3.2	CO_2 and the earth's temperature gradually changed in tandem	24
Figure 3.3	Correlation of recent temperature anomaly with CO_2 rise	27
Figure 3.4	The carbon cycle diagram (anthropogenic data as of mid-1990s)	28
Figure 3.5	What would a gigametric ton look like?	29
Figure 3.6	Visualizing big numbers	29
Figure 3.7	Rise in global average temperature	30
Figure 4.1	Population growth and global energy demand	32
Figure 4.2	U.S. greenhouse gas emissions	33
Figure 4.3	Total U.S. greenhouse gas emissions by economic sector (2016)	34
Figure 4.4	Per capita CO_2 emissions versus GDP per capita	35
Figure 4.5	GDP and CO_2 per capita	36
Figure 4.6	Three countries plus the EU accounted for over half of CO_2 emissions	37
Figure 5.1	Global atmospheric temperature rate of change	41
Figure 5.2	Average global sea surface temperature, 1880−2015	43
Figure 5.3	Average sea level rise	44
Figure 5.4	Greenland ice melt	49
Figure 5.5	Reduction in Greenland ice mass	49
Figure 5.6	Arctic sea ice, winter versus summer	50
Figure 5.7A	Arctic sea ice, 1979	51
Figure 5.7B	Arctic sea ice, 2014	51
Figure 5.8	Deviation in sea ice extent (Mkm^2)	52
Figure 5.9	The Columbia Glacier, Alaska. Left: 2009. Right: 2015	53
Figure 6.1	CO_{2eq} gas components	66
Figure 6.2	Global CO_2 emissions by world region, 2014	67
Figure 6.3	Where do emissions go?	68
Figure 6.4	Global carbon dioxide emissions from fossil fuels	71
Figure 9.1	Global emissions of CO_{2eq} in 2018	108
Figure 9.2	IPCC global warming scenarios and projections	109
Figure 9.3	IPCC scenarios	113
Figure 9.4	Example of overshoot and return to temperature target	113
Figure 9.5	IPCC goal compared to business as usual	116
Figure 9.6	What happens if no action is taken?	119
Figure 10.1	Global energy consumption by fuel	123
Figure 10.2	Global energy consumption by sector	124

Figure 10.3	Mismatch between solar power supply and demand	131
Figure 10.4A	LCOE Wind Power	132
Figure 10.4B	LCOE utility-scale PV solar	132
Figure 10.5	Trends in wind turbine design	133
Figure 10.6	The biofuels life cycle	134
Figure 11.1	Germany's greenhouse gas emissions compared to 1990, in $GmtCO_{2eq}$.	150
Figure 12.1	U.S. public policy priorities 2019	168
Figure 13.1	World oil consumption per capita	183
Figure 14.1	CO_2 emissions declined relative to GDP due to higher efficiency	205
Figure 14.2	Energy efficiency graphs: (A) Refrigerators. (B) LED lamps. (C) Buildings	206
Figure 14.3	Impact of the Action Plan on "Business as Usual"	214
Figure 14.4	Estimated cost of seawall construction in the United States, through 2040	221
Figure 15.1	Levelized cost of energy alternatives in North America	229

List of Tables

Table 4.1	Absolute and per capita greenhouse gas emissions	35
Table 6.1	Greenhouse gas sources	65
Table 9.1	Summary of IPCC scenarios	112
Table 10.1	Fossil fuel types	128
Table 10.2	Biofuels compared to fossil fuels energy content	135
Table 10.3	Renewable fuel alternatives	136
Table 11.1	Germany's greenhouse gas reductions	151
Table 14.1	Action plan summary	213
Table 14.2	Impact of Action Plan on year carbon budget is exceeded	216
Table 14.3	Effect of \$US50/ton of CO_2 carbon fee on fuels	225
Table 15.1	Electrical output from 1 m^2 solar panel	231
Table 15.2	Electrical output from 4 MW wind turbine	232
Table 15.3	Solar and wind power costs including storage	233
Table 15.4	Estimated fossil fuel subsidies	236
Table A.4.1	Key parameters values.	279

xiii

Photographs

Photo 1.1	Coal burning power plant, Iowa	3
Photo 1.2	Mauna Loa Observatory, NOAA, Hawaii	4
Photo 2.1	Wind generating farm, Desert Hot Springs, California	15
Photo 5.1	Polar bear on sea ice, Svalbard Island, Norway	47
Photo 5.2	Sinai desert, Egypt	55
Photo 5.3	Wild fires, California	56
Photo 5.4	Wild fires, California	57
Photo 6.1	Clearing tropics	65
Photo 7.1	Flooding in Newport Beach, California	81
Photo 7.2	Hubbard Glacier, Alaska	85
Photo 7.3	Live Joshua Trees, Joshua Tree National Monument	88
Photo 7.4	Dying Joshua Trees, Joshua Tree National Monument	88
Photo 7.5	Rural Ethiopian village	93
Photo 10.1	Offshore oil platform, Gulf of Mexico	126
Photo 10.2	Micro grid Borrego Springs, California	145
Photo 13.1	Great Plains cornfield	196
Photo 14.1	Subsistence farm Ethiopia	210
Photo 15.1	Solar and wind farms, North Palm Springs, California	228

Preface: Why read this book?

Global warming is probably the most important and complex public policy and international relations issue in the world today. Stopping global warming and mitigating its unavoidable consequences are not just a technical problem. The science and technology available today are sufficient to address this problem without delay. Many aspects of our society will be impacted in some way by global warming. The biggest challenges involve educating the public so that they will support the required changes, getting governments to support a carbon fee and other actions and achieving the level of international cooperation and coordination required to solve this global problem.

The purpose of this book is to provide a complete understanding of global warming with clear explanations of the science behind climate change. The book contains the facts people need to understand global warming and what can and should be done about it. We discuss actions that need to be taken, including a review of relevant past successes and failures with lessons learned. The complexities and challenges of addressing global warming are discussed along with the unique problems of the six countries that are the largest sources of greenhouse gases.

For specialists such as those in earth science or environmental studies, this book is a good overview covering the science, technology, economics, politics, international relations, and other issues involved in actually doing something about global warming. This should help specialists more narrowly focused on specific problems to understand all the problems and choices involved and the actions that must be taken. For academics, engineers, professionals, and concerned citizens, the book tells them what they need to know to understand, discuss, and debate global warming.

This book will be of interest to students in science, engineering, political science, international relations, economics, public health, law, and perhaps other disciplines. For nontechnical students, the book is an understandable and comprehensive presentation of the science and technology they need to understand global warming and what should be done about it.

We see global warming due to increases in greenhouse gas emissions as the fundamental problem. Climate change is a consequence of global warming. Let us not confuse climate change and weather. Weather is what happens regionally over short periods of time as determined by temperature, humidity, wind speed, rain or snowfall, atmospheric pressure, and some other parameters. Climate is the average atmospheric condition over relatively long periods of time, usually 30 years, for a geographic region. Climate change is a broad term that encompasses the more frequent and severe weather events we are experiencing and other effects due to the earth's higher atmospheric and ocean temperatures.

Carbon dioxide (a molecule consisting of one atom of carbon and two of oxygen and abbreviated CO_2) emissions from combustion of fossil fuels are the main cause of global warming. While CO_2 is the major greenhouse gas, amounting to about 76% of the total, there are other gases, such as methane, that also contribute. When the combined effect of all the greenhouse gases is discussed, they are often referred to as "CO_{2eq}," shorthand for carbon dioxide equivalent. The use of fossil fuels to produce power and heat for consumers and industry accounts for most greenhouse gas emissions. However, 25% is due to agriculture and land-use changes, mainly the destruction of forests to produce more land for agriculture.

Only by understanding the problem can intelligent solutions be evaluated and supported by a majority of the world's population. Global warming is further complicated by the fact that severe consequences may not occur for decades. We will not experience the full effects of our actions today due to latency, the delay between cause and effect. Yet, without prompt action NOW, the earth could reach a point where warming becomes irreversible and we lose the ability to stop it. No matter what actions are taken, the effects of global warming will be felt unevenly, hitting some areas harder than others. Energy costs are likely to increase during the transition to renewable sources and fossil fuel use will have to be largely eliminated. Longer-term, it is possible that energy from renewable sources could be less expensive than from fossil fuels, especially if the public health costs of fossil fuel use are considered. Changes in agriculture and land use will also be necessary.

To stop global warming, we need to get to "net zero," meaning essentially eliminate greenhouse gas emissions due to human activity. Anything less will slow down but not prevent ever-increasing temperatures. Even after net zero is achieved, it will take centuries for greenhouse gas concentration to naturally decline to preindustrial levels.

Global warming over the last 100 years or more is due to emissions of greenhouse gases caused by human activities, principally the combustion of fossil fuels to produce power and heat. As greenhouse gases build up in the atmosphere, they reduce the amount of heat leaving the earth, causing it to warm up. Ocean temperatures are increasing, polar ice is melting, permafrost is melting, and the sea level is rising due to the combined effects of thermal expansion and melting ice. Other physical evidence of global warming can be seen, as we describe. The earth has had many cycles of climate change in its history—usually over periods of tens of thousands of years. What makes these recent changes noteworthy (and ominous) is that they are occurring at an unprecedented rate—in decades, not in millennia, as in prehistoric times.

In 2014 the Intergovernmental Panel on Climate Change (IPCC), after detailed studies by a group of world climate experts, set a goal of limiting the earth's average temperature increase to $1.5°C$ ($2.7°F$) above "preindustrial levels." IPCC estimated that meeting this temperature goal would require keeping the atmospheric concentration of CO_2 below 450 parts per million (ppm). As we show in Chapter 3, the highest concentration during the 400,000 years preceding the Industrial Revolution was 300 ppm. By June 2019, the atmospheric concentration had reached a high of 415 ppm.[1] Today, it is certain that the IPCC goal of limiting to a $1.5°C$ temperature rise cannot be met. Greater temperature increases are inevitable. The consequences of higher temperatures are unknown, but are likely to be severe.

The earth's average temperature has already increased $1.0°C$ above its temperature before the start of the Industrial Revolution (estimated as $13.8°C$ or $56.8°F$ in 1880) and is headed toward a $2.0°C$ rise. While $2.0°C$ ($3.6°F$) may seem like nothing, for humans, a $2.0°C$ rise from the normal body temperature of $37°C$ ($98.6°F$) could mean a fever of $39°C$ ($102.2°F$) and a serious illness.[2]

Sadly, it is not good enough to stop the rate of greenhouse gas emissions at current levels. At current emission levels, greenhouse gas concentrations in the atmosphere will continue to increase indefinitely, leading to more global warming. CO_2 released to the atmosphere stays there for hundreds of years. If CO_2 emissions stopped today, it would take hundreds of years for the excess CO_{2eq} already in the atmosphere to completely dissipate. Meanwhile, warming would continue.

Readers are challenged by the large amount of climate disinformation disseminated by the fossil fuel industry and others, and by the complexity of much of the available valid science. Several organizations and individuals minimize the impact of global warming or promote the belief that nothing can or should be done about this problem. Others are promoting naïve and unrealistic solutions or timelines that cannot be met, such as getting to net zero by 2050. This book counters misinformation and misunderstandings concerning global warming.

There are no easy or quick solutions. The task ahead will be difficult, but doable. Nations may procrastinate. The degree of cooperation, foresight, and sacrifice needed may be beyond our current capabilities on a national or global basis. We may have to experience a severe crisis before effective actions are taken. We can see a substantial increase in the earth's temperature with drastic effects on the earth's climate and local weather patterns. *Failure is an option.*

This book addresses the science of global warming in a readable manner and provides references for readers who want more detail or to study the sources of information. It is organized into three parts, 16 chapters, and numerous clearly labeled subchapters listed in the Table of Contents to make it easy for the reader to quickly identify topics of interest. Part 1 discusses the science behind global warming and the impact of global warming on our environment. Part 2 discusses what it would take to get to net zero, the alternatives available, and the unique problems to be solved, with relevant case studies and lessons learned. Part 3 is the recommended action plan to get to net zero with a realistic forecast of what is possible. This part discusses the feasibility of powering the world with renewable energy and the cost of making the transition to eliminate fossil fuel use. The importance of educating the public, implementing a carbon fee, and the need for greater international cooperation are discussed. The book concludes with recommendations for governments, industry, and concerned citizens covering the technical, economic, public policy, and international issues involved, and outlines a positive way forward.

The book proposes practical solutions that can be implemented now with today's technology that will significantly reduce greenhouse gas emissions and put the world on a path toward net zero, the elimination of greenhouse gas emissions. Technical and economic trends that are in the right direction are not moving fast enough. These positive trends need to be accelerated. An optimistic future is possible with abundant energy to maintain rising living standards without most of today's air pollution. The negative consequences of not addressing this problem are covered as well.

Although the IPCC goal of achieving "net zero" by 2050 is unachievable in the authors' opinion, we should start immediately with the technology we have. Addressing the problem now will stimulate new approaches and the development of new technologies needed to get to net zero eventually. We provide a plan to address global warming that includes greater energy efficiency, replacing fossil fuels with renewable solar and wind power and other practical measures. A carbon fee—a price on CO_2 emissions—is needed to offset the massive subsidies for fossil fuel use. Although cash and tax subsidies are significant, the biggest subsidies arise from the ability to discharge pollutants and greenhouse gases into the atmosphere at no charge. A carbon fee will apply a realistic cost to the effects of fossil fuel pollution and greenhouse gas emissions. If fossil fuel subsidies are not offset, the transition to renewable, nonpolluting sources of energy will be delayed.

The most important concepts in the book are illustrated with easy-to-follow calculations using actual, real-world data so that the reader can understand and replicate the analyses if they so desire. This book is a useful reference or handbook for those concerned about global warming. There are extensive references and endnotes that would be helpful for readers looking for additional detail on the topics covered. Useful websites and major reports are listed in the appendices, which also include abbreviations, conversion factors, sample calculations, and other relevant information.

If you want to know more about global warming and what to do about it, read this book. If you like what you read, inform your friends. An informed and concerned citizenry is necessary to address this critical problem.

William D. Fletcher and Craig B. Smith
Newport Beach, CA, United States

End Notes

1. CO_2. earth, Daily CO_2 readings, Mauna Loa Observatory, https://www.co2.earth/daily-co2.
2. With thanks to John Goodman, who suggested this analogy in a letter to the *Los Angeles Times*, December 5, 2018, p. A12.

Acknowledgments

It would not have been possible to write this book without the dedicated efforts of thousands of scientists in many disciplines working today and in the past. Their efforts have gathered, analyzed, and reported the scientific data needed to write this book. They have conducted experiments to verify and explain the changes we are observing. Without their efforts, the world would be experiencing changes without understanding what is happening and without insights concerning what can and should be done to prevent adverse effects. We are all in their debt. We owe thanks to many people who encouraged us to write this book and who offered helpful suggestions and critiques. We are especially indebted to the scientists, engineers, economists, lawyers, and others who reviewed an early draft of the manuscript: Curtis Abdouch, M.S., Captain Jerry Aspland, Cecelia Arzbaecher, Ph.D., William Michael Barnes, Ph.D., John J. Berger, Ph.D., Paul Bjorkholm, Ph.D., John E. Bond, Attorney, Virginia Casey, M.S., Frank E. Coffman, Ph.D., Kevin C. Daly, Ph.D., Jerry Dauderman, MBA, Peter Fletcher, M. S., Suzy Fletcher, M.L.S., Warren Fix, Joe Genshlea, Attorney, Rich Harms, Raymond W. Holdsworth, MBA, Tony Hsu, Ph.D., Marie Kontos, educator, John Kensey, MBA, Dave Larue, Ph.D., Wilbur D. Layman, Attorney, René Malés, M.S., John Martin, BSEE, Tom Merrick, M.S., Phil and Jane Miller, Tom Osborne, Ph.D., Gary Palo, M.S., Kelly Parmenter, Ph.D., James (Walkie) Ray, BSCE, Andrew Smith, Ph.D., Nancy Smith, educator, Robert Smith, Ph.D., Russell Spencer, Ph.D., Mark Taggert, environmental lobbyist, Robert Taylor, environmental writer, Richard Thompson, BSEE, MBA, Mitzi Wells, banker.

A special thanks to Shahir Masri, Ph.D., from the University of California, Irvine, who kindly reviewed the preliminary draft and then reread the final draft for technical accuracy.

We thank Matthew Laffin, Tenaya Parmenter, Halle DeMargo, and Mike Phillips for developing the illustrations, and Raeghan Rebstock for the website and social media design and support. Also thanks to Ms. Valeen Szabo, Director, Borrego Springs Chamber of Commerce, for providing information regarding the Borrego Springs micro grid project.

To those individuals and organizations who generously gave permission to reproduce illustrations from their publications, we express a special thanks: James Balog, Photographer, Earth Vision Institute; Professor

xxiii

Steven Chu, Ph.D., Stanford University; Ms. Judi Mackey, Managing Director, Global Communications, Lazard; Mr. Niall McCarthy and Statista; Ms. Julia O'Hanlon, Pew Research Center; Pinya Sarasas, Programme Officer, UN Environment Program, Nairobi; and Axel Schweiger, Chair, Polar Science Center, University of Washington.

Photos were provided from the personal collections of Bill Fletcher, Craig Smith, and Curt Abdouch.

We gratefully acknowledge the support and encouragement of our wives Suzy Fletcher and Nancy Smith.

We appreciate the professionalism and support of key staff at Elsevier. Lisa Reading, Senior Acquisition Editor, Energy, responded to our initial inquiry and referred us to Marisa LaFleu, who initiated the external review of our proposal. Next, Peter Llewellyn, Acquisitions Editor, encouraged us with his enthusiasm and support for our project, guiding us through the contract process and into the capable hands of Michelle Fisher and Chiara Giglio, Editorial Project Managers, who brought the book to life. Kumar Anbazhagan was Project Manager for overseeing the book through the production process. In addition, Ms. Ashwathi Aravindakshan assisted as Copyright Coordinator and Unni Kannan Ramu helped us with the financial paperwork.

Finally, any errors are our sole responsibility.

Acronyms

AEC	Atomic Energy Commission
CAGR	Compound annual growth rate
DOD	US Department of Defense
DOE	US Department of Energy
EIA	US Energy Information Agency
EPA	US Environmental Protection Agency
EU	European Union
G20	Group of 20
GDP	Gross domestic product
IEA	International Energy Agency
IGY	International geophysical year
IPCC	Intergovernmental Panel on Climate Change
IRENA	International Renewable Energy Agency
LCOE	Levelized cost of energy
NASA	US National Aeronautics and Space Administration
NCA	US National Climate Assessment
NDC	Nationally determined contribution
NOAA	US National Oceanic and Atmospheric Administration
NRC	Nuclear Regulatory Commission
OAPEC	Organization of Arab Petroleum Exporting Countries
OECD	Organization for Economic Cooperation and Development
OPEC	Organization of Petroleum Exporting Countries
UN	United Nations
UNFCC	United Nations Framework Convention
WMO	World Meteorological Organization
WHO	World Health Organization
WRI	World Resources Institute

CHAPTER 1

Introduction

This introductory chapter is for concerned readers who may not have a science background but want to know more about global warming and what to do about it. It includes an overview of the book and presents the key points with a minimum use of figures and analysis.

Throughout the book we use metric units for the data presented. This is because the metric system is the global standard measurement system and because this book is written for a global audience. Where appropriate, we also show units that U.S. readers are more familiar with, such as degrees Fahrenheit (°F) when temperatures are shown as centigrade degrees (°C). The Appendices have conversion tables that can be consulted if needed.

Should we be concerned about global warming?

Global warming is a serious threat to life on earth as we know it and cannot be ignored. Most importantly, due to vested interests by the petroleum, coal, and natural gas businesses, the public has been subjected to conflicting and misleading information, and politicians have refrained from passing meaningful legislation for fear of losing financial or political support. The public hesitates to press their governmental representatives about a matter that seems to be far in the future and that might impact them financially through higher costs for fuel and electricity. To better understand the need for immediate action, we need to recognize that *latency*, or possible catastrophic delayed effects of global warming, is extremely important. Chapter 2, Addressing Global Warming, addresses latency and other challenges and outlines in summary form what needs to be done.

Reaching Net Zero.
DOI: https://doi.org/10.1016/B978-0-12-823366-5.00001-4
© 2020 Elsevier Inc.
All rights reserved.

What about solar radiation?

As we all know, the earth is warmed by solar radiation as it travels around the sun. The earth's orbit is not constant; the earth wobbles, it tilts back and forth a small amount, and the orbit changes shape a bit. Every day the earth receives an excess of energy in the form of solar radiation. Fortunately, some of this energy is reflected by the earth's atmosphere and most of the rest that hits earth is reradiated back into space. Otherwise, the earth would become too warm. CO_2 and other greenhouse gases in the atmosphere are transparent to incoming solar radiation, but trap some of sun's heat on earth that otherwise would go back into space. Over the last several million years, the balance of these two effects—incoming and outgoing solar energy—have kept the earth's temperature in a range where human life is possible.

The greenhouse effect

Most readers are familiar with a greenhouse. The phenomenon is the same as what we experience when an automobile is parked outside on a sunny day. Radiant energy from the sun penetrates the windows and warms the interior of the vehicle. Some of this radiant energy is absorbed by the automobile's interior. The rest attempts to leave the vehicle as radiant energy but is reflected back in by the window glass, heating the interior. This is very important for the earth. Trapping some of the sun's radiant energy warms the earth and makes it habitable. The problem arises when there is too much greenhouse gas in the atmosphere. This is like adding thicker glass to the greenhouse. More heat is trapped and the earth gets hotter. This change began following the Industrial Revolution and is continuing. The earth's average temperature has already increased $1.0°C$ ($1.8°F$) since the start of the Industrial Revolution in England in 1750. Our concern is that unless early and effective actions to reduce CO_2 and other greenhouse gas emissions are taken, the earth's average temperature rise could increase to $2°C$ ($3.6°F$) or more. As pointed out in the Preface, $2°C$ may not seem like much, but to life on earth it would be very significant.

What are the greenhouse gases?

Carbon dioxide is the most common greenhouse gas and accounts for about 76% of greenhouse gases. The next largest greenhouse gas is methane, followed by nitrous oxides and fluorinated gases released by industrial processes. Most of the carbon dioxide, 90%, comes from the burning of fossil fuels, namely coal, oil, and natural gas. Coal- and natural gas-fired power plants generate electricity. Oil-based products such as gasoline, diesel fuel, and aviation fuel provide most of the energy used in transportation. Industry also uses fossil fuel to produce power and heat needed by industrial processes. Residential and commercial buildings use electricity for air-conditioning and lighting and oil and natural gas for heating (Photo 1.1).

Land-use changes, mainly the destruction of forests to clear land for crops and animals, are another source of carbon dioxide. Trees and other plant materials absorb carbon dioxide as part of photosynthesis and naturally remove carbon dioxide from the atmosphere. When forests are destroyed, more carbon dioxide has to remain in the atmosphere or is absorbed by the oceans.

Photo 1.1 Coal burning power plant, Iowa.

Why is carbon dioxide suddenly a problem? For a long time, probably at least 100 years, few people thought anything about the fact that human beings were now burning increasing amounts of coal and then oil and natural gas. We were too busy enjoying the benefits provided by abundant low-cost energy.

In 1958 a young atmospheric scientist employed by Scripps Institution of Oceanography named Charles Keeling began making measurements from an observatory on the top of Mauna Loa, on the Big Island of Hawaii. This site was selected because it was in the middle of the Pacific Ocean and relatively unaffected by air pollution and other effects from the continents. After several years of measurements, Keeling discovered that the concentration of carbon dioxide in the atmosphere was steadily increasing. He subsequently devoted his life to continuing these measurements and was followed later by his son and other research institutions. As a result, we now know that there has been an ever-increasing concentration of carbon dioxide in the atmosphere. Other measurements indicate that the average temperature of the earth has been increasing simultaneously with the increase in carbon dioxide. In effect, adding carbon dioxide to the atmosphere is equivalent to adding more layers of glass to the greenhouse, causing the earth's temperature to rise (Photo 1.2).

Photo 1.2 Mauna Loa Observatory, NOAA, Hawaii.

What are the signs of global warming?

Other than the fact that the earth's average temperature has suddenly started to increase at an unusually rapid pace, what other symptoms of global warming have been observed? For one, sea levels are rising. This is due to the fact that oceans are warmer (water expands when heated) and due to melting of glaciers and polar ice. In the northern latitudes, permafrost is also melting, with the potential to release trapped methane, a potent greenhouse gas. Deserts are expanding and droughts are becoming more frequent. Extreme weather events are occurring more often.

What are the dangers of global warming?

A major concern is that earth's temperature will continue to rise. Global warming can be compared to driving an automobile with a sticky accelerator and no brakes. We expect the automobile to start to slow down as soon as we take our foot off the accelerator. If the car does not slow down fast enough, we can step on the brakes to slow it down or even bring it to a complete stop.

Global warming does not work that way. When carbon is released to the atmosphere, it stays there for hundreds of years. The temperature will not go down for a long time after greenhouse gas emissions cease. It is as if we took our foot off the accelerator, but the automobile will not slow down. We have essentially put thicker glass in our "greenhouse" that perpetuates global warming until slow-acting natural processes gradually remove carbon dioxide from the atmosphere over hundreds of years.

We can step on the accelerator, emit more greenhouse gases, and increase global warming but we cannot slow it down once the greenhouse gases have been put into the atmosphere. What about brakes? The equivalent of brakes would be the ability to actually remove greenhouse gases from the atmosphere. At present, there is no practical or economical process to do this other than planting billions of trees to expand the earth's forests. Even if artificial methods for removing carbon dioxide become practical, it would take huge amounts of energy and be very expensive, to remove enough carbon dioxide to actually lower the earth's temperature.

Other known risks include increasing air pollution, rising sea levels that will cause coastal damage and flooding, and ocean warming that will affect weather, fish populations, and coral. As more carbon dioxide dissolves in the oceans, they become more acidic, affecting shellfish and coral. A very serious risk is the melting of glaciers, polar ice caps, and sea ice, not only because of rising sea levels but also because of the potential reduction in freshwater supplies. As deserts expand and the earth continues to warm, high temperature zones will tend to move northward. This will cause plant and animal species to migrate in some cases and to become extinct in others. Insects and diseases will migrate and there will be impacts on agriculture and human health. Humanitarian crises will occur where food supplies and access to freshwater are reduced or temperatures become too hot for humans. This will lead to people migrating to other areas looking for food, water, jobs, and a more hospitable climate (Textbox 1.1).

TEXTBOX 1.1: Key points of this book.

- Global warming is man-made and a threat that has to be addressed without further delays.
- Fossil fuels are an outdated solution that needs to be replaced by renewables as quickly as possible.
- Major trends are in the right direction but not moving fast enough, such as the increased use of renewable energy and electric vehicles. These trends should be accelerated.
- There is a risk that we could reach a tipping point that will accelerate global warming in a potentially irreversible way, beyond our control. This is not a prediction but a possibility that should not be ignored.
- We cannot get to net zero fast enough to meet the Intergovernmental Panel on Climate Change's goal of keeping global warming under 2°C. The current forecast is for warming to exceed 3.0°C by 2100.
- Due to latency, climate change will continue even if greenhouse gas emissions are reduced to net zero.
- Due to ongoing climate changes, mitigation has to be a big part of any plan to address global warming.
- Relevant lessons learned are summarized from the successes and failures of past and present large programs, such as the development of

(Continued)

Introduction 7

TEXTBOX 1.1: (Continued)

commercial nuclear power in the United States and Germany's Energiewende program to phase out coal and nuclear power.

- Austerity need not be part of the solution. A positive future is discussed with abundant and affordable energy from renewable sources.
- An achievable plan is presented to power the world with renewable energy. Specific bottlenecks are identified and solutions proposed. Most air pollution is eliminated in addition to greenhouse gases.
- The economics of making the transition to renewables is analyzed. Considering all the costs and benefits involved, funds should be available to make the transition.
- The transition to renewable energy and changes needed to combat global warming should present business opportunities to those who embrace change. China is presently ahead of the European Union and the United States and is capturing these business opportunities.
- If we do not do the big things, the small things may be commendable but are not sufficient to stop global warming. The big things are: (1) educate the voting public so that they will support needed changes, (2) institute a carbon emissions fee to offset fossil fuel subsidies, and (3) participate in international efforts to stop global warming.
- A planned smooth transition to renewables is needed to avoid major recessions, malinvestments, stranded assets, and uninsured losses. A disruptive transition is likely to compromise public support. Major infrastructure projects and siting and land use guidelines should be included in any plan.
- Failure is an option.

Can anything be done about global warming?

First, we need to recognize that with greenhouse gases, what goes into the atmosphere stays in the atmosphere. Carbon dioxide will take a long time—centuries—to dissipate naturally. Reducing greenhouse gas emissions is not enough. Reducing emissions will delay but not stop global warming. Think of it this way. I can slowly accelerate this car up to 70 miles/hour. But unless I do something, it keeps going. It is the same with the carbon dioxide already in the atmosphere. It will go on *increasing* the earth's temperature until we take action. To slow global warming, we have to eliminate the use of fossil fuels and make other changes, mainly related to land use. Otherwise, temperatures will keep rising.

To answer the question about what needs to be done—first we have to stop using fossil fuels. One good recommendation is to charge coal, oil, and natural gas companies a fee based on the quantity of carbon that comes from their mines or wells. It will create an incentive to leave fossil fuels in the ground and to accelerate the switch to cleaner forms of energy. That money can be used to combat pollution, partly offset rising energy prices, and to build solar and wind power plants and associated infrastructure such as transmission lines. Wherever possible, we need to use renewable sources of energy, mainly electricity generated by solar and wind but also hydro and nuclear power. In transportation, electric vehicles need to replace those powered by internal combustion engines to the extent practical. There will be some fossil fuel uses that cannot be eliminated. These uses will need to be satisfied using hydrogen or synthetic fuels derived from recycled CO_2 produced using electricity from renewable sources.

There may be a way to suck carbon dioxide out of the atmosphere and store it back underground, but so far it has not been done on a large scale and may never be practical. It may be technically feasible but uneconomical. We need to continue to develop carbon removal technologies to determine their technical feasibility and costs. We also need to preserve and expand the world's forests which naturally absorb CO_2 via photosynthesis.

The next chapter is a summary of what we believe must be done to reduce global warming and details the challenges that are involved. It also has a guide or "roadmap" to the balance of the book for readers who might want to skip some of the technical parts and go directly to the action plan and way forward.

CHAPTER 2

Addressing global warming

Global warming is a real and present danger to the earth as we know it. If we do not deal with global warming now with the technology we have today, we will not be able to maintain the world's population and our standard of living. The authors' objective is not to write another book about the causes and dangers of global warming. We do cover these topics in the interest of providing the reader with a complete discussion of this subject. Our main purpose is to:
- Explore the challenges associated with taking action on global warming. This will be an extremely difficult problem to solve for many reasons.
- Present a plan of action, a way forward. It will take an estimated 50 years or more to significantly reduce global warming if we start soon. There are no quick solutions. Stopping greenhouse gas emissions will not stop global warming immediately, but will slow future temperature increases.

What is the challenge? According to the October 2018 special report by the Intergovernmental Panel on Climate Change (IPCC), to limit global warming to less than 1.5°C, greenhouse gas emissions would have to be reduced to net zero by 2050.[1] *This means that any man-made emissions would have to be completely eliminated (brought to "net zero") or offset by "sinks" that absorb CO_2, the main greenhouse gas. These sinks could be natural, such as forests and other plant life that take up CO_2, or by as yet unproven carbon removal and storage systems. Other greenhouse gases such as methane would have to be reduced to net zero as well.*

Our forecast is that global warming will exceed 2.0°C even if we start to take action today. Considering the practical problems discussed in this book, the United States and the rest of the world are unlikely to reduce fossil fuel use fast enough and make other changes needed to get to net

---

[1] V. Masson-Delmotte, et al., *Global Warming of 1.5° C. An IPCC Special Report on the Impacts of Global Warming of 1.5° C above Pre-industrial Levels and Related Global Greenhouse Gas Emission Pathways, in the Context of Strengthening the Global Response to the Threat of Climate Change, Sustainable Development, and Efforts to Eradicate Poverty*, IPCCSR1.5 (Geneva, Switzerland: World Meteorological Organization, 2018), https://www.ipcc.ch/sr15/.

zero by 2050. Nonetheless, we have to start immediately with the technology we have today. As we describe in Chapter 5, How Do We Know Global Warming Is Real?, and Chapter 7, What Are the Effects of Global Warming?, some damage has already been done and more will occur even if we take action now.

This problem will not solve itself. We can take the initiative and start tackling this problem now or wait until we are forced to take action in the future when there is some global warming disaster that is too big to ignore. The longer we wait, the harder the problem is to solve. There will be more greenhouse gases in the atmosphere, our fossil fuel use will have increased, and there will be more damage that has to be dealt with.

The sun provides much more solar energy each day than is required to meet all the world's energy needs. This energy is available to everyone and every country. The challenge is to harness this clean source of energy using nonpolluting devices such as solar cells and wind turbines to power the global economy and meet our individual energy needs. *Fortunately, this is practical today and increasingly cost-effective using the tools and technology we have.* What we have is sufficient to substantially reduce greenhouse gas emissions even if we are not sure how to get to net zero by 2050, the goal recommended by the IPCC.

In November 2018, the U.S. Government issued the *Climate Science Special Report: Fourth National Climate Assessment (NCA4)*, a collaborative effort involving 12 federal agencies.[2] *This assessment concludes, based on extensive evidence, that it is extremely likely that human activities, especially emissions of greenhouse gases, are the dominant cause of the observed warming since the mid-20th century. For the warming over the last century, there is no convincing alternative explanation supported by the extent of the observational evidence.* This report also states that without major reductions in emissions, the increase in annual average global temperature relative to preindustrial times could reach $5°C$ ($9°F$) or more by the end of this century.

Many people are unaware of the solid science supporting the concern about global warming. There is uncontroversial physical evidence for those who have an open mind. Past temperature cycles occurred over *thousands* of years and have been within a fairly narrow range. Recently, the earth's temperature has risen very rapidly (in decades, rather than millennia) compared to past cycles. Temperature increases closely follow

---

[2] Wuebbles et. al., *Fourth National Climate Assessment (NCA4)*, U.S. Climate Science Special Report (U.S. Global Change Research Program, 2018), https://www.globalchange.gov/nca4.

changes in greenhouse gas concentrations in the atmosphere. By June 2019, the peak CO_2 concentration reached 415 parts per million (ppm) compared to a maximum of about 300 ppm over the past 400,000 years.[3] The greenhouse effect, the primary mechanism by which increasing greenhouse gases warm the earth, is a well-researched phenomenon supported by scientific experiments.

Some people confuse weather with climate change. How can there be global warming when we are having such a cold winter, for example. As noted in the Preface, weather is short-term changes in temperature, rainfall, wind, and other weather effects. Weather is highly variable. Climate is the average weather over a long period. Climate change also leads to changes in the weather for specific locations. It is not unusual for some areas to become colder or wetter, for example, while other areas become warmer.

Taking action on global warming involves some big, difficult, and initially expensive changes. If we, for example, are not willing to transition away from fossil fuel use, then the smaller stuff will not make much of a difference. We can already see problems with the public's opposition to a carbon tax or higher gasoline taxes. Buying an electric car will not help much if the electricity used is produced by burning coal or natural gas in a power plant. Adding attic insulation or switching to energy-efficient light bulbs is commendable but insufficient to make a significant contribution to solving this problem.

Latency is a huge problem

The effects of today's greenhouse gas emissions will not be fully realized for a long time, perhaps 25, 50, or even 100 years from now due to latency, the delay between cause and effect. Professor Steven Chu, a Nobel Laureate and former U.S. Energy Secretary, used the delay between cigarette consumption and the incidence of lung cancer to illustrate the effect of latency. Lung cancer deaths lagged increased cigarette consumption by about 25 years, the typical time required for a smoker to

---

[3] NASA, Global Climate Change, Vital Signs of the Planet, Graphic: The Relentless Rise of Carbon Dioxide, site last updated January 9, 2020, https://climate.nasa.gov/climate_resources/24/graphic-the-relentless-rise-of-carbon-dioxide/. Shows preindustrial CO_2 levels were 300 ppm or less.

develop lung cancer (Fig. 2.1). For global warming, greenhouse gases put into the atmosphere stay there for a very long time, causing the earth's temperature to remain elevated. The earth will be hotter, ice will continue to melt, and other effects will persist indefinitely. Natural processes take a very long time to reduce atmospheric greenhouse gas concentrations and will not have a measurable effect until long after greenhouse gas emissions are eliminated.

Global warming is not obvious to the vast majority of people

Now people are beginning to experience the effects of global warming—extreme heat, droughts, forest fires, air pollution, storms, and flooding—that are visible. But because these events have occurred in the past, they are not thinking "climate change" as much as unusual weather, such as 100-year storms. The connection is missing. Think about smog that you could see and feel in Los Angeles in the 1960s or in Beijing or New Delhi today—we are familiar with these events. In contrast, while we may read about rising sea levels, heat waves, melting ice caps and glaciers, or forest fires, it is hard to get excited about something we are not experiencing personally. An exception might be people who have been subject to weather extremes during the recent past. These include longer droughts, larger and more frequent wildfires, new heat records, and more severe storms. Still, many people who have lost their homes due to a wildfire or flooding do not seem to connect the loss to global warming.

The global economy is powered by fossil fuels

For more than 200 years or so, we have depended upon power and heat from fossil fuels to provide the comfort and conveniences we enjoy today. The production and use of fossil fuels are a huge global business that employs millions. It will be very difficult to make a transition away from fossil fuel use voluntarily. A rapid transition is not practical. It is likely to take 50 years or more to reach the IPCC's target of net zero. This essentially requires the elimination of most fossil fuel uses. Any

Addressing global warming 13

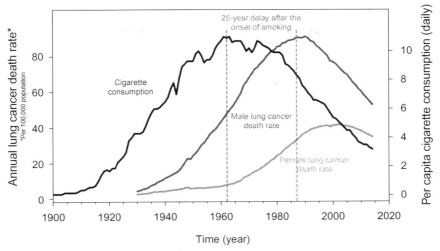

Figure 2.1 An example of latency.[4]

residual fossil fuel use needs to be offset by natural sinks such as expanded forests that absorb CO_2 naturally. Artificial means of removing CO_2 from the atmosphere are under development but may not be practical or economical.

There is a need for unprecedented and perhaps unachievable international cooperation

All the world's countries share a common biosphere (atmosphere, oceans, plant and animal life). If some countries effectively deal with global warming while others do less, then the problem will not be solved. It is especially important that the United States, China, and the other five or six largest emitters of greenhouse gases work together to solve global warming no matter what their differences are in other areas.

---

[4] Steven Chu, Ph.D., "Energy and Climate Change: Challenges and Opportunities," presentation to Stanford Business School, Stanford University, Palo Alto, June 4, 2014. Chu cites cigarette smoking and lung cancer as an example of latency. Used with permission.

All of us will have to be willing to accept changes

This is a problem that cannot be solved without all of us agreeing to make some changes on behalf of stopping global warming. The biggest sacrifice is the need to pay more for the fossil fuels we use directly and indirectly, and to eventually eliminate fossil fuel use. The effects of global warming will be uneven. Will people be willing to make changes to avoid flooding in some distant location such as Bangladesh, or even Florida? Will the current generation make sacrifices so that future generations can have a better life?

U.S. participation is essential

The U.S. cannot expect other countries to do more than it is willing to do. The United States must remain a member of the Paris Agreement (*L' Accord de Paris*) and demonstrate leadership in dealing with global warming. The United States is the world's second largest emitter of greenhouse gases after China, a country with four times the U.S. population. On a per capita basis, the United States is the largest greenhouse gas emitter with the exception of Canada, Saudi Arabia, and some other much smaller economies. The United States accounts for the largest share of CO_2 in the atmosphere today, an estimated 25%. This is because the United States has been a major emitter of greenhouse gases since about 1860, when the industrial revolution began in the United States.

What can be done?

What can we do about global warming? The obvious answer is to use alternatives to fossil fuels and make the transition as soon as we can. This will reduce CO_2 emissions. We also need to reduce the emissions of other greenhouse gases, mainly methane and nitrous oxides, and also increase carbon sinks to absorb any CO_2 emissions that cannot be eliminated. This will require major changes in land use and agriculture, such as preserving forests and planting millions of new trees (Photo 2.1).

Photo 2.1 Wind generating farm, Desert Hot Springs, California.

Fortunately, solutions are available. We can meet most of the world's energy needs using renewable energy from solar cells and wind turbines. The use of these sources of energy is increasing rapidly, but not fast enough. Electric energy from solar and wind is steadily declining in cost and is increasingly competitive with fossil fuels and nuclear power. Some fossil fuel uses will persist. So, yes, there are practical problems to be solved—but we cannot let that prevent us from getting started now. There are problems to be solved along the way, but they are not showstoppers that prevent us from starting.

Part of the solution is a carbon emissions fee. This should not be viewed as a tax. It is reasonable to charge a fee for the discharge of CO_2 and other greenhouse gases to compensate for the damage done by global warming and air pollution and the cost to mitigate the effects of these emissions. To the extent that CO_2 and other greenhouse gas emissions are emitted at no cost, we are giving fossil fuels a huge subsidy compared to other energy sources. Revenues from a carbon fee are also needed by governments to fund essential programs dealing with global warming. Without these revenues, spending is likely to add to government deficits and debt.

Importantly, fossil fuels will only stay in the ground when electricity and liquid fuels from renewable sources are cheaper and as reliable as the fossil fuels they displace. A carbon fee would increase the cost differential in favor of renewables.

What do we have to do? To stop global warming we need to do five things:

1. Transition to renewable sources, solar and wind, for the production of electricity. We especially need to reduce dependence on coal. Coal is the dirtiest fossil fuel and the primary fuel used for electricity production worldwide.

2. Increase overall energy efficiency. Energy efficiency, the amount of economic output as measured by GDP per amount of energy used, is already increasing by about 2.0%/year. We propose that measures be taken to increase energy efficiency by an additional 1.0%/year. Between now and 2050, this would reduce the world's energy use by another 30%. This additional 1.0% is readily achievable with expanded use of LED lighting, use of heat pumps and more efficient appliances, sustainable building design, and more efficient industrial processes including microwave heating. In addition, solar and wind avoid the thermodynamic inefficiencies that result when electricity and power are generated from fossil fuels (no energy lost as waste heat from combustion).

Addressing global warming

3. For activities that use fossil fuels, convert as many as possible to electricity as their source of power and heat. Basically, electrify the economy. The biggest challenge is to convert transportation systems to electric or hydrogen-powered vehicles. Over time, we need to abandon the internal combustion engine.
4. Improve agriculture to reduce methane and nitrous oxide emissions. Changes in agriculture can also absorb and sequester large amounts of CO_2.
5. Change land-use practices, mainly deforestation, to preserve and preferably expand the ability of forests and other plants to absorb CO_2. Plant billions of trees.

Can we afford to do these things? Can we afford not to? The possible costs and benefits associated with the elimination of fossil fuels for most uses over the next 50 years or so are numerous. For example, there will be a big health bonus associated with the reduction and eventual elimination of fossil fuel use. Fossil fuels are the source of almost all of the world's air pollution. This air pollution also contaminates our water supplies and food supplies, as some pollutants fall to earth. In many locations, air pollution is a serious public health problem. The World Health Organization estimates that seven million people die annually from air pollution.[5]

This book is organized in parts so that readers who are already convinced that global warming is real can proceed to later chapters. *Part one* (Chapters 3–8) covers global warming basics. What is global warming and what are the likely effects of global warming? *Part two* (Chapters 9–13) covers current efforts to slow global warming, has some useful case studies, and tells why this is such a difficult problem to solve. *Part three* (Chapters 14–16) are the authors' recommended action plan—what we need to do with a realistic forecast of possible reductions over time, and how we should proceed to address global warming.

What should the reader do? Do more than read this book. Discuss global warming with others and help educate the public. Within the general public, there is an incredible amount of misinformation and complacency about global warming, the dangers it presents, and what we can and should do about it. Educating the public and instilling a sense of urgency to deal with global warming is perhaps our biggest challenge. Your efforts can help.

---

[5] World Health Organization (WHO), Air Pollution, updated 2018, https://www.who.int/airpollution/data/en/. WHO estimates 7 million deaths per year from air pollution.

PART I

CHAPTER 3

The earth as a system

To a scientist or an engineer, the earth is a very large system. It is dynamic but stable. By dynamic, we mean that key variables are ever changing. By stable, we mean that changes oscillate around stable values such as temperature and a predictable orbit in space.

In the most elementary view, the earth is a sphere that circles the sun in an elliptical orbit once each year. This reality has important consequences for our climate.

Incident solar radiation

The source of energy for the earth is radiation from the sun. Each *day*, the earth receives an estimated $15,000 \times 10^9$ gigajoules (GJ) of energy in the form of radiation. Currently the world's population uses about 600×10^9 GJ per *year*. Thus if the incoming solar energy could be fully converted to human use, one day's worth would provide the energy needs of the earth's population for about *25 years*. It is a lot of energy, but we receive it unevenly—more at the equator, less at the poles, and on only half of the planet at any given time. (Textbox 3.1).

One would think that receiving all this energy every day would cause the earth to heat up. It would, if there were not offsetting processes at

> **TEXTBOX 3.1 About those pesky international units.**
> Joule (J) is the energy needed to produce 1 watt of power for 1 second.
> A gigajoule (GJ) is a billion joules.
> *Note*: A joule is a small unit. A candy bar (250 calories) has the equivalent energy of about 1 million joules. A British Thermal Unit (Btu) is 1054 Joules.
> Degree centigrade (°C) = 1.8 degrees Fahrenheit (°F).
> Water freezes at 0°C and boils at 100°C.
> Metric ton (mt) is 1000 kilograms or 2204 pounds.

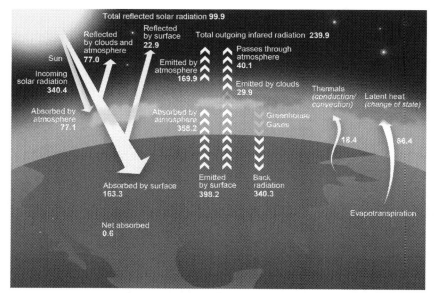

Figure 3.1 Solar energy sources and sinks.[1]

work. Part of the energy the earth receives, a small part, is used by living plants to convert the sun's radiant energy into plant material via photosynthesis. Photosynthesis takes the energy from the sun, CO_2 in the atmosphere, and nutrients absorbed from the ground to grow trees, grasses, crops, and other plants. Another portion of the incoming solar radiation is absorbed in the land, lakes, and oceans.

The incoming solar energy is short-wavelength radiation that includes the visible light that we see. The visible spectrum is centered around a wavelength of 0.5 microns. The vast majority of the sun's radiant energy is not used on earth but is radiated back into space as long-wavelength infrared radiation centered around a wavelength of 15 microns. See Fig. 3.1 for solar energy sources and sinks.

Although the earth is obviously much cooler than the sun, it is much warmer than the empty space that surrounds it. For an asteroid traveling through space, the earth is a hot object that radiates energy back into

---

[1] "Solar sources and sinks," National Aeronautics and Space Administration (NASA). Source: Norman G. Loeb, et al., "Toward Optimal Closure of the Earth's Top-of-Atmosphere Radiation Budget," *Journal Climatology* 22 (2009): 748–766 and Kevin E. Trenberth, John T. Fasullo, and Jeffrey Kiehl, "Earth's Global Energy Budget," *Bulletin of the American Meteorological Society (BAMS)* (March 2009): 311–324.

space. Earth's average temperature at present is estimated to be 15°C (59°F), while the temperature of space is −270°C or (−454°F), a difference of 285°C.

Over the eons, the world has been hotter and much cooler than it is today. There have been ice ages and times when there was very little ice, when the sea level rose hundreds of feet, and the land turned tropical. However, as noted above, these natural changes to the earth's climate took a long time—tens of thousands of years per cycle. How do we explain these large changes in the earth's climate?

Milankovitch cycles

Changes in the earth's climate from warm periods to ice ages are due to changes in the amount of radiant energy the earth absorbs from the sun. There are three major perturbations to the earth's orbit around the sun. They are known as the Milankovitch cycles.[2] They have an important bearing on climate. The first is *precession*. The earth spins with a slight wobble. It completes one complete wobble roughly every 26,000 years. The earth's axis is tilted, currently about 23.5 degrees from the vertical. This *tilt* is of great importance since it causes the seasons. The tilt fluctuates back and forth a few degrees, taking about 41,000 years to fully complete a cycle. Finally, the earth's orbit also varies *obliquely*, from elliptical to more circular, with a cycle that takes about 100,000 years. The combination of these three effects causes the earth to heat up or cool down. This is a very slow process that takes about 100,000 years for a full cycle as shown in Fig. 3.2. This is very different from the much more rapid changes in the earth's temperature that have occurred in the last 150 years or so.

We can also see from Fig. 3.2 that the concentration of CO_2 in the atmosphere has been closely followed by changes in the earth's temperature. This close relationship has held for at least the last 800,000 years, which is as far back as sophisticated ice core measuring techniques have enabled us to look. This is another example of the earth being a dynamic (changing) system, but stable, cycling within long-established boundaries

---

[2] Shahir Masri, *Beyond Debate: Answers to 50 Misconceptions on Climate Change* (Newport Beach, CA: Dockside Sailing Press, 2018), 3−5.

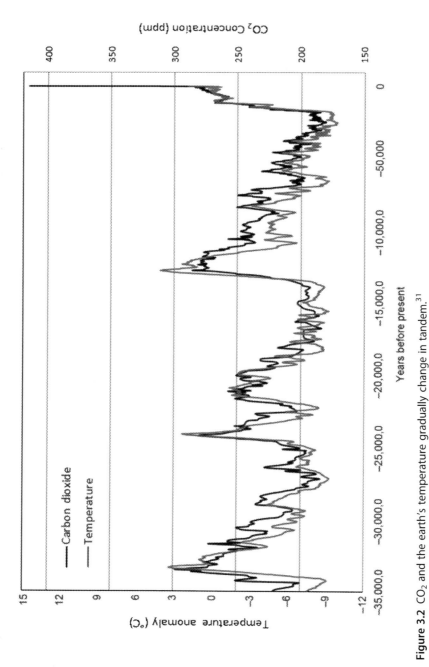

Figure 3.2 CO_2 and the earth's temperature gradually change in tandem.[31]

[3] NASA, Global Climate Change, Graphic: The Relentless Rise of Carbon Dioxide, site last updated January 9, 2020, https://climate.nasa.gov/climate_resources/24/graphic-the-relentless-rise-of-carbon-dioxide/. Shows CO_2 variation in ancient times.

The earth as a system

and maintaining relationships between variables, in this case CO_2 concentration and temperature.

More about the greenhouse effect

Today, the earth's temperature is no longer primarily driven by the Milankovitch cycles, but by the greenhouse effect. What we are seeing is that man-made (anthropogenic) changes are increasing the greenhouse effect and causing near-term increases in the earth's temperature. Human activity, mainly the use of fossil fuels, is overriding natural forces, leading to rapid increases in the earth's temperature. Were it left to natural cycles, the Earth would in fact be cooling, as was slowly happening for thousands of years. Yet, today, human activity has entirely reversed that trend, now pushing us into dangerous warming. The destruction of forests and land-use changes due to agriculture also contribute. Temperature increases that usually take thousands of years are now happening in decades and are moving in the opposite direction that natural cycles would dictate.

Fig. 3.2 also shows that the concentration of CO_2 in the atmosphere did not exceed about 280 ppm until the start of the Industrial Revolution. In June 2019, the peak CO_2 concentration in the atmosphere reached 415 ppm, about 50% higher and rising. In addition, the earth's average temperature has increased in proportion to the increasing CO_2 concentration—an increase of $1.0°C$ (range is estimated to be $0.83°C-1.16°C$) since about 1850. This increase in CO_2 in the atmosphere and the increase in the earth's average temperature are occurring too rapidly to be caused by the Milankovitch cycles. There are no other natural forces that can explain these increases.

How can CO_2 be a problem? The problem is that CO_2 is a "greenhouse gas" and has a powerful effect on the amount of the sun's incoming radiation that is radiated back into space. A greenhouse maintains a higher temperature than its surrounding area because some of the radiation entering the greenhouse is prevented from leaving. The short-wavelength radiation entering the greenhouse penetrates the glass. But, radiation attempting to leave the greenhouse has a longer wavelength and a portion is reflected back, warming the interior. The phenomenon is the same as that described in Chapter 1, Introduction, when entering an automobile parked outside on a sunny day.

The greenhouse effect is essential for life on earth. The CO_2 in the atmosphere has trapped enough of the sun's incoming energy to keep the planet warm enough to support life, with an average earth temperature of 15°C (59°F). Without some naturally occurring greenhouse gases, the earth's average temperature would be a chilly (−)18°C (0.4°F), below freezing.[4] But this is a delicate balance, a balance that has been upset to the point that the earth is now warming.

Since the start of the industrial revolution in the U.S. around 1850, humans have been burning increasing amounts of wood, coal, oil, and natural gas to produce the energy needed to cook food, heat homes, power industry, produce electricity, and power transportation (planes, trains, trucks, cars, ships). The Industrial Revolution started in England much earlier, in 1750. However, England's economy was much smaller than that of the U.S., and England's early contribution to global warming was much less. Greenhouse gas emissions and their effects are discussed in more detail in Chapter 5, How Do We Know Global Warming Is Real?, and Chapter 6, How Do We Know Man-Made CO_2 Is the Issue?.

Carbon cycle

As noted in Fig. 3.3, since 1880, the amount of CO_2 in the earth's atmosphere has steadily increased from 280 to 410 ppm (annual average value) in a short period of slightly more than 150 years. This sharp rise is what distinguishes the current era from ancient times. It has been accompanied by a sudden increase in the earth's average temperature. "Temperature anomaly" refers to the *difference* in current temperature from the historical baseline. The carbon rise is significant, because an increase of just 1.0 ppm in CO_2 concentration is equivalent to 7.86 *more* Gmt of CO_2 in the atmosphere, a mass equivalent to that of about a *billion* elephants (see Fig. 3.5).

CO_2 in the atmosphere is part of the earth's carbon cycle. There is a large natural flow of carbon between several reservoirs. Over the long

---

[4] Qiancheng Ma, "Greenhouse Gases: Refining the Role of Carbon Dioxide," NASA, Goddard Institute for Space Studies, Science Briefs, March 1998, https://www.giss.nasa.gov/research/briefs/ma_01/. Shows that, without CO_2, earth's temperature would be −18°C.

The earth as a system

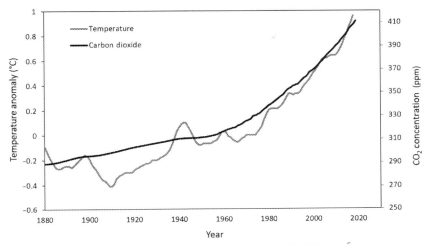

Figure 3.3 Correlation of recent temperature anomaly with CO_2 rise.[5]

term, carbon flow between reservoirs is in balance. Changes that put more CO_2 into the atmospheric reservoir increase the earth's temperature by the greenhouse effect.

There are two carbon cycles, a slow cycle and a fast cycle. The slow cycle is the process by which carbon is transformed into rocks in the form of limestone (calcium carbonate) or into shell-building organisms that eventually form coral reefs or are captured in sediment at the bottom of the ocean. Coal, oil, and natural gas are forms of carbon that were plant materials several million years ago and were transformed over time by geological forces into the fossil fuels used today.

The fast carbon cycle is largely the flow of carbon through life forms on earth or in the oceans. This flow is seasonal. Less CO_2 is emitted during the growing season when plants absorb more CO_2 and more is emitted during other times. It is a massive annual flow of carbon. Imbalances in the fast carbon cycle due to man-made emissions of CO_2 are the cause of today's global warming (Fig. 3.4).

Fig. 3.4 shows a representation of the carbon cycle. The solid lines represent the flow of carbon (C, not CO_2) during the preindustrial age, while the dashed lines show the effect of man-made emissions. The solid lines show 119.6 + 70.6 or 190.2 GmtC/year entering the atmosphere and a like

---

[5] Climate Central, "The Globe is Already Above 1°C, on Its Way to 1.5°C," October 9, 2018, https://www.climatecentral.org/gallery/graphics/the-globe-is-already-above-1c.

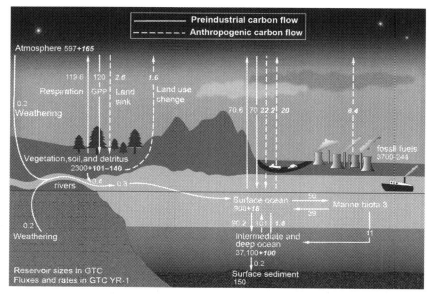

Figure 3.4 The carbon cycle diagram (anthropogenic data as of mid-1990s).[6]

> **TEXTBOX 3.2 Note about GmtC.**
> *Note:* GmtC is shorthand for "Giga metric tons of carbon" or one billion metric tons of carbon.

amount returning to the earth. Not so with the dashed lines. There is a net 3.2 GmtC/year more than is absorbed on earth going into the atmosphere. [This was the International Panel on Climate Change (IPCC) estimate for the mid-1990s; today the amount is much greater] (Textbox 3.2).

To better visualize this massive quantity, it might help to realize that one gigametric ton is equivalent to the weight of 150 *million* male African elephants (see Fig. 3.5). One gigametric ton is the amount of CO_2 dumped into the atmosphere every month by China and every 2 months by the United States.

Throughout this book, it is necessary to refer to some very large numbers. Fig. 3.6 shows a visual representation of large numbers to help the

---

[6] Based on global carbon cycle in IPCC 2007, *The Fourth Assessment Report*, https://www.ipcc.ch/assessment-report/ar4/. See Working Group 1 portion of its Fourth Assessment Report (Figure 7.3 in Chapter 7: "Couplings Between Changes in the Climate System and Biogeochemistry"). The values for human influences represent the state of the carbon cycle in the mid-1990s.

The earth as a system 29

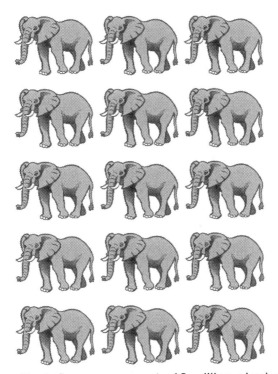

Each figure represents 10 million elephants!
(1 Gmt is equivalent to the weight of 150,000,000 elephants)

Figure 3.5 What would a gigametric ton look like?

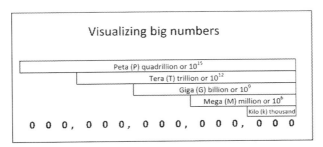

Figure 3.6 Visualizing big numbers.

reader. Using a number followed by 10 to a power is referred to as scientific notation. The power of 10 refers to the number of zeros following the number. For example, one Giga is a billion written as 1 followed by nine zeros, 1,000,000,000. See Appendix 1 for additional information on units and conversion factors.

Temperature increase

The average increase in the earth's temperature is now *at least* 1.0°C (1.8°F) higher than the IPCC baseline. It is estimated that the earth's temperature is increasing about 0.2°C (0.36°F) every decade (see Fig. 3.7). This seemingly small change potentially has serious consequences. The increase may be accelerating.

The IPCC uses the average temperature for 1850−1900 of 13.7°C (56.7°F) as a baseline as shown in Fig. 3.7. Using this baseline, the earth's average temperature has increased 1.0°C (0.8°C−1.16°C). This book will use the IPCC baseline throughout. The reader may come across another estimate from the National Oceanic and Atmospheric Administration (NOAA) that uses the earth's average temperature from 1951 to 1980 as a baseline. Using this baseline, the earth's temperature increase is about 0.8°C. However, both these estimates agree on the current temperature of the earth, 15°C (59°F).

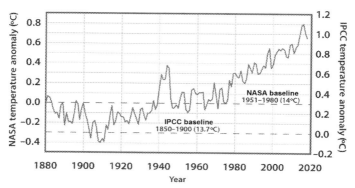

Figure 3.7 Rise in global average temperature.[7]

---

[7] NASA, Goddard Institute for Space Studies, GISS Surface Temperature Analysis (v4), https://data.giss.nasa.gov/gistemp/zonal_means/index_v4.html. NASA data for average global temperature.

CHAPTER 4

Fundamental drivers of global warming

There are two fundamental forces driving global warming. The first is the world's large and increasing population. The second is rising living standards throughout the world. As people's living standards improve, they use more energy.

Global population rising

Today, the world's population is 7.6 billion people, up from about 1.1 billion in 1860, the start of the industrial revolution in the United States. The world's population is increasing about 1.1%/year. The United Nations forecasts that the world's population will reach about 9.9 billion by 2050 and 11.1 billion by 2100. All these people will need a lot of energy, as indicated in Fig. 4.1.

Inequities: the early role of the United States and the United Kingdom

There are inequities that need to be addressed in any discussion of the causes of global warming. The United States is the world's largest industrial economy and started its industrialization in about 1860, before other countries. The United Kingdom was the first country to industrialize but is a much smaller economy than the United States. The growth in fossil fuel use began in 1750, following the onset of the Industrial Revolution in the United Kingdom. This development was paralleled in the United States beginning about 1860. A little-acknowledged fact is that the United Kingdom and the United States "own" a significant part of the CO_2 currently in the atmosphere.

How could this be? From 1853 to 2016, the United Kingdom produced 25,448 million metric tons of coal and imported 924 million metric

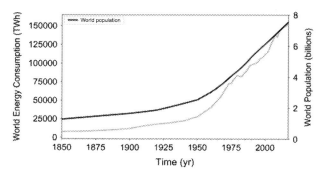

Figure 4.1 Population growth and global energy demand.[1]

tons, for a total of 26,372 million metric tons. By the year 2016, coal production was virtually negligible in the United Kingdom. Before 1975 there was virtually no oil production in the United Kingdom.

During the period 1800–2017, the United States produced 76,410 million metric tons of coal. In addition to coal, the United States produced 222 billion barrels of oil from 1850 to 2017. In 1900 the United Kingdom and the United States alone emitted 1.3 Gmt of CO_2, or about two-thirds of the world's total of 2 Gmt of CO_2 emitted at that time. As of 2016, global cumulative emissions were 26% for the United States, 13% for China, 5% for the United Kingdom, and 3% for India, the leading contributors.[2]

---

[1] Sources: Our *World in Data.org*. Population data may be found at: https://ourworldindata.org/grapher/world-population-by-world-regions-post-1820 and is from the United Nations, Department of Economic and Social Affairs, Population Division (2019). World Population Prospects: The 2019 Revision, DVD Edition. Available at: https://esa.un.org/unpd/wpp/Download/Standard/Population/. Energy data may be found at: https://ourworldindata.org/grapher/global-primary-energy and is derived from two sources: Appendix A of Vaclav Smil's Updated and Revised Edition of his book, 'Energy Transitions: Global and National Perspectives' (2017), and BP Statistical Review of World Energy. All data prior to the year 1965 is sourced from Smil (2017). All data from 1965 onwards, with the exception of traditional biomass is sourced from BP Statistical Review. Smil's estimates of traditional biomass are only available until 2015. For the years 2016 onwards, we have assumed a similar level of traditional biomass consumption. This is approximately in line with recent trends in traditional biomass from Smil's data. *Our World in Data* has normalized all BP fossil fuels data to terawatt-hours (TWh) using a conversion factor of 11.63 to convert from million metric tons of oil equivalent (Mtoe) to TWh. Data represents primary energy (rather than final energy) consumption. See following links: http://vaclavsmil.com/2016/12/14/energy-transitions-global-and-national-perspectives-second-expanded-and-updated-edition/https://www.bp.com/en/global/corporate/energy-economics/statistical-review-of-world-energy.html

[2] Hannah Ritchie and Max Roser, "CO_2 and Greenhouse Gas Emissions," Our World in Data, last revised December 2019, https://ourworldindata.org/co2-and-other-greenhouse-gas-emissions. Shows the long-run history of cumulative CO_2.

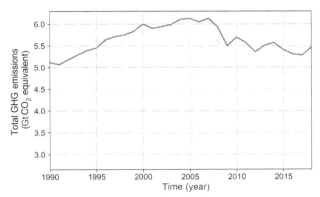

Figure 4.2 U.S. greenhouse gas emissions.[5]

The pivotal position of the United States

Fig. 4.2 shows *total* U.S. greenhouse gas (GHG) emissions, [CO_2 plus CH_4 (methane), NOx, hydrofluorocarbons, and perfluorinated compounds (PFCs)]. The total declined from a peak of 6511 million metric tons of CO_{2eq} in 2005 as natural gas began to replace coal as a power plant fuel to produce electricity. What are the sources of these emissions? Fig. 4.3 shows the breakdown between transportation, electricity generation, commercial, residential, and agriculture uses.[3] Offsetting these emissions is land and forests (not shown in the graph), which act as a sink, absorbing CO_2 from the atmosphere. In the United States, since 1990, managed forests and other lands have absorbed more CO_2 than they emit.[4] On a global basis, land-use changes are net emitters of CO_2.

---

[3] U.S. Environmental Protection Agency, Sources of Greenhouse Gas Emissions, accessed January 13, 2020, https://www.epa.gov/ghgemissions/sources-greenhouse-gas-emissions. Shows break down of total U.S. GHG emissions by economic sector in 2017.
[4] Gabriel Popkin, "How Much Can Forests Fight Climate Change?," *Nature* 565 (2019): 280–282. https://www.nature.com/articles/d41586-019-00122-z.
[5] U.S. Environmental Protection Agency, *Inventory of U.S. Greenhouse Gas Emissions and Sinks: 1990–2015*, EPA 430-P-17-001, April 15, 2017, https://www.epa.gov/sites/production/files/2017-02/documents/2017_complete_report.pdf. See Fig. ES-1 for total U.S. GHG emissions by year.

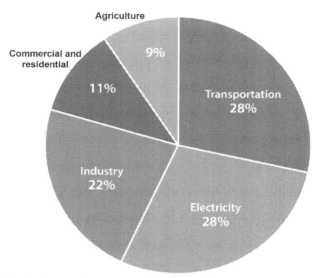

Figure 4.3 Total U.S. greenhouse gas emissions by economic sector (2016).[6]

 ## Need to consider both absolute and per capita emissions

Another point that is often ignored is per capita CO_{2eq} emissions. Total CO_{2eq} emissions by country do not tell the whole story. Per capita emission of CO_{2eq} is a more meaningful statistic. As seen in Table 4.1, the biggest absolute emissions come from China because of its large population, followed by the United States and India. On a per capita basis, the United States is the largest emitter of the major countries. China ranks 47th on a per capita basis. Based upon total emissions, India is the third largest country but only 158th on a per capita basis due to India's large population but low standard of living. We can see that China's per capita CO_2 production is less than half that of the United States. India's is only about 10% of the United States. On a per capita basis, U.S. CO_2 production is about 3.3 times the global average (see Fig. 4.4).

Currently, about 85% of the world's energy is provided by the combustion of fossil fuels: coal, oil, and natural gas. As living standards increase, people need more energy for everything from electric lighting, transportation, heating and air conditioning, communications, and to

---

[6] Ibid. See page 2–23 for U.S. GHG emissions by economic sector.

Table 4.1 Absolute and per capita greenhouse gas emissions.[7]

Country	CO₂ yield (Mmt/year)	Population (millions)	mtCO₂ per capita
United States	5172	326	16.1
Russian Federation	1760	136	12.3
Japan	1252	128	9.9
Germany	778	82	9.6
China	10,641	1390	7.7
Brazil	486	209	2.3
India	2455	1260	1.9
Nigeria	86.9	161	0.5
Ethiopia	10	97	0.1
Global average	36,061	7400	4.9
Very underdeveloped countries[a]	Very low	3400	Negligible

[a] According to the World Bank, more than 3.4 billion people live in poverty defined as living on less than $5.50/day. This is almost half the world's population. These people use very little energy and do not contribute much to greenhouse gas emissions via fossil fuel consumption.

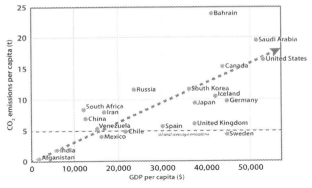

Figure 4.4 Per capita CO₂ emissions versus GDP per capita.[8]

drive industrial production (see Fig. 4.4). Fossil fuels are also used as raw materials to produce essential products such as plastics and fertilizers. Energy use is closely correlated with GHG emissions. The biggest energy users are also the biggest emitters of GHGs.

---

[7] Marilena Muntean, et al., *Fossil CO₂ Emissions of all World Countries — 2018 Report* (Publications Office of the European Union, 2018), doi:10.2760/30158, https://ec.europa.eu/jrc/en/publication/fossil-co2-emissions-all-world-countries-2018-report.

[8] European Environmental Agency, Correlation of Energy Consumption and GDP per Person, last modified July 7, 2016, https://www.eea.europa.eu/data-and-maps/figures/correlation-of-per-capita-energy.

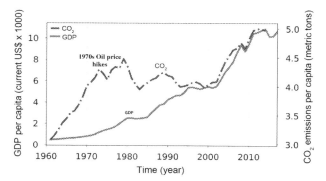

Figure 4.5 GDP and CO_2 per capita.[9]

Gross domestic product growth and energy use are related

For the 20-year period following the Arab Oil Embargo and its sharp increase in oil prices, per capita CO_2 emissions remained relatively flat, due to global efforts to increase energy conservation and efficiency in response to the higher cost of fuel (see Fig. 4.5). If pre-1970s oil prices had continued, emissions would be much greater today. After about 20 years of flat per capita energy use, from 1979 to 2000, we see that per capita CO_2 emissions again increased in parallel with increasing gross domestic product (GDP) as oil prices declined due to supply increases. The long-term effects of the 1970s oil price hikes clearly show that higher energy prices lead to conservation and efficiency improvements. This is discussed in more detail in Chapter 13, Some Success and Failures.

Reducing global energy consumption is an important measure for reducing CO_2 emissions. But, it is unlikely that it can be effective as the primary means of combating global warming. Conservation and improving energy efficiency will help. However, the only effective way to combat global warming is to replace fossil fuels with renewable forms as the primary source of energy. While the combustion of fossil fuels is the main source of GHGs,

---

[9] Sources: GDP per capita data from: The World Bank, GDP per Capita (Current US$), World Bank and OECD National Accounts Data, accessed January 13, 2020, https://data.worldbank.org/indicator/NY.GDP.PCAP.CD. CO_2 data from: Carbon Dioxide Information Analysis Center, Climate and Environmental Sciences Division, Oak Ridge National Laboratory, last modified September 11, 2017, https://cdiac.ess-dive.lbl.gov/home.html.

Fundamental drivers of global warming

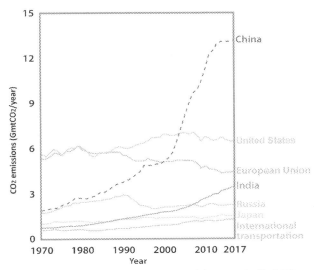

Figure 4.6 Three countries plus the EU accounted for over half of CO_2 emissions.[10]

there are other sources that should also receive attention, from agriculture to land-use changes such as the practice of clearing forests to create cropland.

 ## More energy will be required by developing countries

We can see from Fig. 4.6 the rapid growth in GHG emissions from China, and to a lesser extent India, as these countries take steps to raise living standards for very large populations. China has the world's largest population, 1.4 billion, but a low growth rate of 0.6%/year due to China's one-child policy, but China's birth rate has not increased much now that more families live in urban environments and living standards are increasing. This policy was modified in 2015 to be a two-child policy. India's population is about 1.3 billion and the growth rate is 1.2%/year. India's population is forecast to exceed China's by 2024.

China and India are examples of a number of countries that will experience growing energy demand as living standards improve.

---

[10] See UN Environmental Programme, 2018 Gap Report, Figure 2.3 in Final report pdf, https://www.unenvironment.org/resources/emissions-gap-report-2018. Used with permission.

CHAPTER 5

How do we know global warming is real?

Global warming is not a new idea

The earliest practical greenhouses were built in Europe in the 1800s. Efforts had been made, even in Roman times, to devise ways to protect plants from adverse weather. Enclosing plants in structures with glass walls and roofs became possible in England and Holland when glass became inexpensive. These greenhouses enabled solar radiation to penetrate and warm the contents while reducing heat loss. Around this same time, various scientists became interested in what caused the ice ages. In 1824 mathematician Joseph Fourier coined the term "Greenhouse Effect" to describe the phenomenon thought to be responsible. Experiments led to the discovery that some gases, in particular CO_2, were not transparent to infrared radiation (longer wavelengths). Sunlight is incident on the Earth in the visible spectrum (light of shorter wavelength), but is reradiated back into space as infrared. (Refer back to Chapter 3: The Earth as a System and Fig. 3.1 to visualize this.) This suggested the possibility that lowering atmospheric CO_2 levels could have increased the heat lost from the Earth and caused the ice ages.

It was left to Svante Arrhenius, a Swedish Nobel Prize winner, to carry out the analysis that proved this proposition in 1896. His calculations showed that if the concentration of atmospheric CO_2 declined by 50%, the global temperature would decrease by 4°C. But not to worry, he suggested, given the amount of coal the world is burning, the planet would stay warm. He calculated that doubling the amount of CO_2 in the atmosphere would increase the global temperature by 4°C. Prophetic words indeed.

Jumping ahead 120 years, how do we know that what Arrhenius prophesied so long ago is *actually happening*? Remember, he did not even have a computer when he made his calculations. A lot of extremely talented people all over the world have been carefully studying this issue for the last 50 years or so, and with them being scientists with no particular ax to grind, it makes you pay attention when they all seem to be

Reaching Net Zero.
DOI: https://doi.org/10.1016/B978-0-12-823366-5.00005-1

© 2020 Elsevier Inc.
All rights reserved.

39

coming to the same conclusions. Here is a summary of what they are reporting.

CO$_2$ emissions are rising

As noted in Fig. 3.2, for hundreds of millennia the concentration of CO_2 varied by about 100 parts/million (ppm), ranging from a low of 180 ppm to a high of 300 ppm. With the advent of the Industrial Revolution, this changed.

Atmospheric concentrations of CO_2 have been measured since 1958 at Mauna Loa in Hawaii and are known for earlier periods by deep ice cores obtained from drilling in Antarctic or Greenland ice sheets. CO_2 concentration in air bubbles trapped in the ice is measured. For 800,000 years before the industrial revolution, CO_2 concentration was in the range of 180−280 ppm. These records show a sharp increase beginning abruptly following the Industrial Revolution, when the combustion of fossil fuels underwent a rapid expansion. Since 1958 CO_2 increased by 24%, first by 2 ppm/year, then by 2.5 ppm/year, and most recently at 3.0 ppm/year.[1] The rate of increase is accelerating. In 2019 CO_2 peaked at 415 ppm.

Earth's temperature is rising

The average temperature of the Earth has gone up and down for millions of years. By measuring the fraction of oxygen isotope O^{18} in deep-sea sediments, and from deep ice-core samples, estimates have been made for the temperatures of earlier times. These estimates have to be regarded as approximate, but indicate periods when the average Earth's temperature (with no polar ice) was as much as $14°C$ ($25°F$) above the 1960−90 average, or as much as $5°C$ ($9°F$) lower during the ice age. From the geologic record these cycles occurred slowly over long periods of time, typically 40,000−100,000 years.

---

[1] NASA, Global Climate Change, Global Temperature, NASA's Goddard Institute for Space Studies Global Land-Ocean Temperature Index, site last updated January 9, 2020, https://climate.nasa.gov/vital-signs/global-temperature/. Shows average land temperature anomaly relative to 1951−80 average. Note: these data use a different (higher) baseline than IPCC.

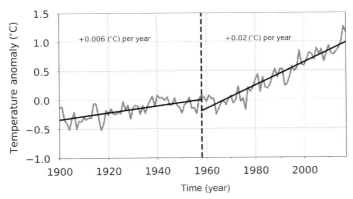

Figure 5.1 Global atmospheric temperature rate of change.[3]

There are actual physical measurements for more recent times. From 1860 to 1980, there are instrument data (thermometers) that have been interpolated to create a global temperature history. From 1980 to the present, we have satellite temperature data, confirmed by ground measurements. These show small oscillations in temperature, as noted in Fig. 3.7. This graph illustrates the change in global surface temperature relative to the 1850–1900 average temperatures. The 10 warmest years in the 136-year record all have occurred since 2000, with the exception of 1998. The year 2016 ranks as the warmest on record.[2] The year 2019 was the second warmest ever recorded in the last decade. This change is measured in decades, whereas in the past the Earth's temperature took thousands of years to change. Note that the rate of change is increasing (see Fig. 5.1). *This rapid change is unusual and is a reason for concern.* We may not be able to slow it down for a long time.

Correlation of increasing global temperature with increasing atmospheric CO_2

The sudden sharp rise in average global temperature parallels the sudden sharp increase in atmospheric CO_2 concentration. There is a

---

[2] NASA, "NASA, NOAA Data Show 2016 Warmest Year on Record Globally," Release 17-006, January 18, 2017, https://www.nasa.gov/press-release/nasa-noaa-data-show-2016-warmest-year-on-record-globally.

[3] James Hansen, Makiko Sato, and Martin Medina-Elizade, "Global Temperature Change," *Proceedings of the National Academy of Sciences of the United States of America* 103, no. 39 (2006): 14288–14293. Article shows the global rate of temperature change is increasing.

consensus among climatologists that the two effects are closely linked, as shown in Fig. 3.3. Note: once discharged into the atmosphere, CO_2 is very slowly absorbed by plants, the oceans, and the Earth. Even if CO_2 emissions stop completely, it will take hundreds of years for the atmospheric concentration to decline, and the Earth's temperature will remain elevated.[4] In other words, greenhouse gases that go in the atmosphere stay there for a long time.

Ocean temperatures are rising

If the planet is warming, one would expect that sea temperature would rise. This is in fact happening, not everywhere, but in most oceans. Changes in temperature affect the oceans in several ways. As water warms, it expands, causing the sea level to rise. The great currents in the ocean follow complex paths, acting as conveyor belts to transport both warm and cold water around the oceans. The Gulf Stream transports warm water from the tropics to the Arctic and cold water from the Arctic to the tropics. This current also warms the ocean and atmosphere along the northwest coast of Europe creating a milder climate there. Changes in water density due to rising temperatures affect these currents. Also, carbon storage is affected by ocean temperature. As oceans warm, they are less able to absorb atmospheric CO_2.

The surface temperature of the sea increased during the 20th century and continues to rise. According to the National Oceanic and Atmospheric Administration (NOAA), from 1901 through 2015, sea surface temperature rose at an average rate of 0.07°C (0.13°F) per decade. Compared to a baseline of the average temperature for the 30-year period from 1971 to 2000, the temperature has risen by 0.6°C (1°F). It has been consistently higher during the past three decades than at any other time since reliable observations began in 1880 (see Fig. 5.2). Warmer ocean water means more energy for storms since hurricanes and typhoons absorb their energy and their moisture from the ocean. Rising ocean temperature also affects fish habitat and migrations. As fish seek cooler waters, other species that depend on them for food follow.

[4] Susan Solomon, Gian-Kasper Plattner, Reto Knutti, and Pierre Friedlingstein, "Irreversible Climate Change Due to Carbon Dioxide Emissions," *Proceedings of the National Academy of Sciences of the United States of America* 106, no. 6 (2009): 1704—1709.

How do we know global warming is real?   43

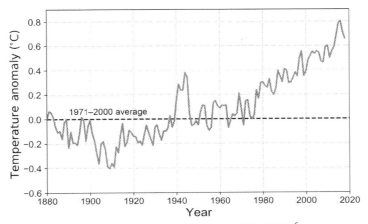

Figure 5.2 Average global sea surface temperature, 1880–2015.[6]

According to the latest information, 2018 was the hottest year on record for the oceans. Commencing in 1990, the rate of ocean warming increased due to the enormous amount of heat being absorbed by the seas as greenhouse gas emissions continue to rise.[5] About 90% of the heat from global warming is absorbed by the oceans and only about 10% is absorbed by the atmosphere.

Sea levels are rising

In September 2019, the IPCC Special Report on the Ocean and Cryosphere in a Changing Climate commented that "the ocean is warmer, more acidic, and less productive. Melting glaciers and ice sheets are causing sea level rise, and coastal extreme events are becoming more severe."[7]

Since the mid-1800s, the average sea level has steadily risen by an average of about 25 cm (10 in.) (see Fig. 5.3). As mentioned in the preceding section, thermal expansion is occurring. In addition, the melting of polar ice caps is also raising the sea level. The effects are already being felt in

---

[5] Lijing Cheng, et al., "2018 Continues Record Global Ocean Warming," *Advances in Atmospheric Sciences* 36, no. 3 (2019): 249–252.

[6] NOAA, Global Climate Report—August 2015, https://www.ncdc.noaa.gov/sotc/global201508/. Shows average global sea surface temperature rise.

[7] H.-O. Portner, et al., eds., *Special Report on the Ocean and Cryosphere in a Changing Climate* (Geneva, Switzerland: IPCC, 2019), https://www.ipcc.ch/srocc/.

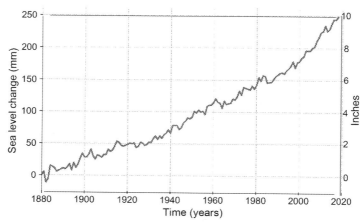

Figure 5.3 Average sea level rise.[11]

low-lying islands and coastal areas around the world. The evidence includes increased coastal flooding and saltwater intrusion in groundwater. As the sea rises, storms and high tides increase coastal flooding and erosion, ultimately destroying homes and crops in low-lying areas. According to NASA satellite measurements, sea level rise is averaging 3.3 mm/year.[8] This rate would lead to an additional increase of 100 mm (0.1 m or 4 in.) by 2050. Near Miami, sea level buoys have measured an 8-in. rise since 1950. The sea level there is now rising at a rate of 1 in. every 3 years.[9] The state is anticipating spending $4 billion to protect sewage systems, raise roads, make stormwater improvements, and build or improve seawalls. These levels are conservative. Depending on the rate at which polar ice caps melt, the sea level rise could be much higher. The IPCC Fifth Assessment report indicates a range of 0.15–0.3 m (6–12 in.) by 2050.[10] With extreme melting of Greenland ice, sea level rise could reach 2.4 m (8 ft.) by 2100.

---

[8] NASA, Global Climate Change, Sea Level, September 2019, https://climate.nasa.gov/vital-signs/sea-level/. Shows average global sea level rise.

[9] Sea Level Rise.org, "Florida's Sea Level is Rising and It's Costing Over $4 Billion," accessed January 13, 2020, https://sealevelrise.org/states/florida/. Shows sea level rise in Florida.

[10] V. Masson-Delmotte, et al., *Global Warming of 1.5° C. An IPCC Special Report on the Impacts of Global Warming of 1.5° C above Pre-industrial Levels and Related Global Greenhouse Gas Emission Pathways, in the Context of Strengthening the Global Response to the Threat of Climate Change, Sustainable Development, and Efforts to Eradicate Poverty*, IPCCSR1.5 (Geneva, Switzerland: World Meteorological Organization, 2018), https://www.ipcc.ch/sr15/.

[11] NASA, Global Climate Change, Sea Level, September 2019, https://climate.nasa.gov/vital-signs/sea-level/. Ground Data: 1870–2013 chart shows average sea level rise since 1880, figure credit: CSIRO, http://www.cmar.csiro.au/.

Ocean acidification is occurring

Measurements indicate that the acidity of the oceans has also increased over the same timeframe as the atmospheric increase of CO_2. About 30% of man-made CO_2 is absorbed by the oceans. CO_2 is converted to carbonic acid when it is dissolved in the ocean. As acidity increases, pH decreases. (Note: pH stands for "Potential of Hydrogen" and is a numeric scale used to specify the acidity or basicity of an aqueous solution.) For example, areas in the Atlantic Ocean and near Hawaii show a downward trend of pH since around 1990.

A message from the Arctic?

Is the Arctic (and the Antarctic) trying to send us a message about increasing global warming?[12] Based on recent measurements, global warming is proceeding twice as fast in the Arctic and Antarctic compared to the global average. This is accelerating the loss of ice at the poles. In the Arctic, permafrost is melting and releasing methane. Hotter and dryer conditions are resulting in forest fires in Alaska, Canada, Siberia, and even Greenland.

The cause of this more rapid warming at the poles is under investigation. According to NASA, a reason for the more rapidly rising temperatures is large weather systems that transport heat to the poles from lower latitudes (approaching the Equator).[13] The Earth gains and loses heat unevenly. The Arctic and Antarctic regions act as heat sinks for the Earth and lose more heat to space than they absorb from the sun. The reverse is true in the lower latitudes where the Earth absorbs more heat than it loses. Ocean currents such as the Gulf Stream and air currents transport excess heat from the lower latitudes to the poles where excess heat is radiated into space.

When temperatures at the poles increase faster than in the lower latitudes, the temperature differential powering ocean currents and air currents to flow toward the poles is decreased. In the northern hemisphere,

---

[12] Jonathan Watts, "Welcome to the Fastest-Heating Place on Earth," *The Guardian*, July 1, 2019, https://www.realclearpolitics.com/2019/07/01/welcome_to_the_fastest-heating_place_on_earth_479107.html.

[13] NASA, Global Climate Change, "NASA Studies an Unusual Arctic Warming Event," site last updated January 9, 2020, https://climate.nasa.gov/climate_resources/164/nasa-studies-an-unusual-arctic-warming-event/.

the jet stream becomes less stable, leading to weather changes in the northern hemisphere. Changes in the Arctic affect heat waves, droughts, wildfires, rainfall, and flooding in the northern hemisphere.

Dramatic evidence of a warmer Arctic is the opening of the Northwest Passage. The Northwest Passage was previously a legendary route connecting the Atlantic and Pacific Oceans by crossing over the top of North America through the ice-bound Arctic Ocean. This route is now sufficiently ice-free during the summer months to allow large ships to take this route.

In Alaska, winter and summer temperatures are rising. Frozen rivers are starting to flow earlier in the season. The Tanana River was ice-free in early April, the earliest in its 103-year record. A big concern is the loss of sea ice in the Bering Sea. Sea ice protects the coastline from winter storms. According to the Alaskan Center for Climate Assessment and Policy, "The most alarming of all the current signs of climate change in Alaska is open water where there should be sea ice."[14]

The melting permafrost is releasing methane, a greenhouse gas, and damaging roads and buildings as the ground shifts. Another problem is that wildfires are more frequent and more severe across the Arctic, not just in Alaska. Similar changes are occurring in Siberia, where methane released by melting permafrost is fueling forest fires and making them more severe. Greenland has hardly any trees. It is largely covered by an ice cap. Greenland's fires are in peat bogs and melted permafrost on the edge of the island.

Svalbard is a group of islands north of the Arctic Circle, halfway between Norway and the North Pole. Longyearbyen is the largest town with a population of about 2100 permanent residents. It is a good example of the fact that Arctic temperatures are rising over twice as fast as the Earth's temperature overall, an estimated 4.0°C so far. The permafrost is melting, leading to landslides, the destruction of buildings, and even forcing the relocation of the cemetery. The climate changes experienced in Svalbard are evidence that the Arctic is experiencing warmer and wetter weather. More precipitation is falling as rain than snow. It is estimated that Svalbard has lost as much as 2 months of winter weather per year, so far. (Photo 5.1).

---

[14] Ian Livingston, "In Alaska, Climate Change is Showing Signs of Disrupting Everyday Life," *The Washington Post*, May 8, 2019, https://www.washingtonpost.com/weather/2019/05/08/alaska-climate-change-is-showing-increasing-signs-disrupting-everyday-life/.

Photo 5.1 Polar bear on sea ice, Svalbard Island, Norway.

Glaciers, ice caps, and sea ice are melting

Greenland is covered by a vast ice sheet, a few meters thick at the fringes but 3200 m (10,500 ft.) thick at its highest point. This ice is supported on land, and if it melted completely, it would raise the sea level worldwide by as much as 20 ft. Recent trends indicate that the Greenland ice sheet is melting more rapidly than would be expected. As noted in Fig. 5.4, Greenland ice builds in the winter and melts in the summer. Fig. 5.5 shows how the Greenland ice mass has decreased from 2007 to 2017.

Satellites revealed that some (but not all) polar ice has been shrinking and glaciers are retreating.[15] The polar ice caps (North and South) contain about two-thirds of all freshwater on Earth. They are different. The Arctic (North Pole) is actually freshwater ice, floating on the Arctic Ocean, called *sea ice*.

Fig. 5.6 shows recent data for the Arctic sea ice; these values are well below the averages of the last 37 years (1979—2016). In 2017 the volume ranged from about 5000 to 20,000 km^3 from summer to winter. Since this ice is floating, if it all melted there would not be an increase in sea level because the floating ice is already displacing an equivalent volume of water.

---

[15] Figure 6: NASA, Global Climate Change, Vital Signs, Carbon Dioxide, site last updated January 9, 2020, https://climate.nasa.gov/vital-signs. Note: This website has a number of useful articles and charts concerning climate change.

Unlike the Arctic, Antarctica is a landmass much larger than Australia and about the size of the Continental United States and Mexico combined. The Antarctica ice sheet has an average thickness of about 2.2 km and covers 98% of the Antarctica continent. It is the world's largest single mass of ice and contains about 90% of the world's freshwater. If this ice sheet were to melt, it would raise sea levels by about 60 m.

The Antarctica ice sheet has several ice shelves or thick layers of ice that extend into the oceans surrounding Antarctica. The largest ice shelf is the Ross Sea Ice Shelf, which is about the size of France and is several hundred meters thick. The ice shelves act as protective buffers that retard the flow of the land ice into the sea. These ice shelves collapse due to surface melting and from melting and erosion of their submerged undersides exposed to warmer ocean temperatures. As ice shelves collapse, the flow of land ice into the sea increases.

A section of the Larson C Ice Shelf on the northern tip of the peninsula, closest to South America, recently broke off. This is the third section that has broken off, following sections "A" and "B." The break created one of the largest icebergs ever, a mammoth the size of Delaware, with an area of 2200 square miles, over 600-ft. thick and weighing a trillion tons. A fissure more than 110 miles long preceded the break. Icebergs are formed when the edges of ice sheets break off and collapse into the ocean, a process called "calving." What makes Larson C unusual is its size. The ice shelves act as protective buffers to retard the flow of land ice into the sea. Ice shelf collapse is driven by surface melt and melting of the lower ice layer, thinning the ice shelf. Warmer temperatures cause more melt and warmer ocean temperatures. The loss of more ice shelves could hasten the melting of land ice and that *would* raise the sea level.[16]

As mentioned above, sea ice is ice supported by the sea rather than by land. Fig. 5.7A and B shows how the sea ice in the fabled Northwest Passage, the treacherous ice-bound sea passage between the Arctic and Canada, is shrinking. Large vessels are now able to make this passage. Fig. 5.7A shows the sea ice in 1979, while Fig. 5.7B shows the same area in 2014. As the Arctic sea ice melts, more of the Arctic Ocean is exposed to the sun, the sea becomes warmer, and ice melts faster. Also, melting ice

---

[16] Sabrina Shankman, "Trillion-Ton, Delaware-Size Iceberg Breaks Off Antarctica's Larsen C Ice Shelf," *Inside Climate News*, July 12, 2017, https://insideclimatenews.org/news/12072017/antarctica-larsen-c-ice-shelf-breaks-giant-iceberg.

How do we know global warming is real? 49

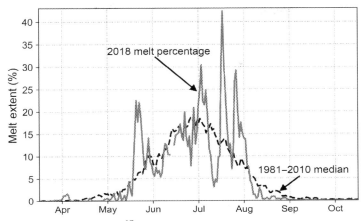

Figure 5.4 Greenland ice melt.[17]

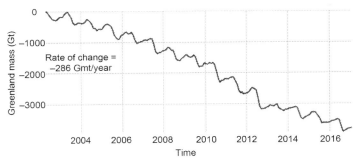

Figure 5.5 Reduction in Greenland ice mass.[18]

means less sunlight is reflected back into space by the darker ocean, contributing to further global warming.

As we can see in Fig. 5.8, the extent of sea ice in the Arctic has declined steadily since 1980. This loss of ice is consistent with increasing Arctic temperatures. Until recently, sea ice in the Antarctic has actually been increasing slowly, reaching its maximum extent in 2014 as shown in Fig. 5.8. Since 2014, Antarctic sea ice has dropped dramatically. Now

---

[17] National Snow and Ice Data Center, Greenland Ice Sheet Today, Greenland Surface Melt Extent Interactive Chart, August 2017, https://nsidc.org/greenland-today/greenland-surface-melt-extent-interactive-chart/.

[18] NASA, Jet Propulsion Laboratory, GRACE Mission Measures Global Ice Mass Changes, accessed January 13, 2020, https://gracefo.jpl.nasa.gov/resources/16/grace-mission-measures-global-ice-mass-changes/. Visualization shows reduction in Greenland ice mass from January 2004 through June 2014. Data are from ice mass measurements by GRACE satellites.

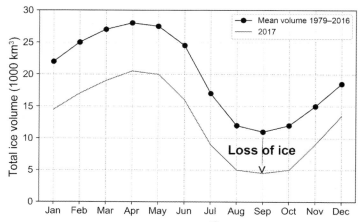

Figure 5.6 Arctic sea ice, winter versus summer.[19]

both the Antarctic and the Arctic are losing sea ice. Today, the extent of sea ice in the Antarctic is the lowest recorded in the 40 years during which sea ice measurements were made.

At present, the causes of the dramatic loss of Antarctic sea ice since 2014 are not well understood. It is not known if this is a temporary phenomenon that may slow down or the start of an ominous trend.

There are roughly 200,000 glaciers on Earth. Rainfall and local temperature variations cause seasonal changes in ice mass. However, overall the general trend has been for glaciers to shrink. A new study shows that glaciers are losing over 300 billion tons of snow and ice each year. What is ominous about this is that they are shrinking five times faster now than in the 1960s. Glaciers in all regions of the world started losing mass around the same time during the past 30 years. The new data, based on more accurate satellite measurements and ground measurements on 19,000 glaciers, also indicate that melting glaciers have a larger impact on sea level rise than previously thought.[20]

---

[19] Polar Science Center, PIOMASS Arctic Sea Ice Volume Reanalysis, Arctic Sea Ice Volume Anomaly, Fig. 2: Total Arctic Sea Ice Volume from PIOMAS showing the Volume of the Mean Annual Cycle, and from 2011 to 2019), accessed January 13, 2020, http://psc.apl.uw.edu/research/projects/arctic-sea-ice-volume-anomaly/. Shows comparison of summer versus winter months. Used with permission from Axel Schweiger, Chair, Polar Science Center, University of Washington.

[20] Associated Press, "Glaciers shrinking faster than thought, study says," p. A8, *Los Angeles Times*, April 9, 2019. See also: World glacier monitoring service, https://wgms.ch/latest-glacier-mass-balance-data/.

How do we know global warming is real?   51

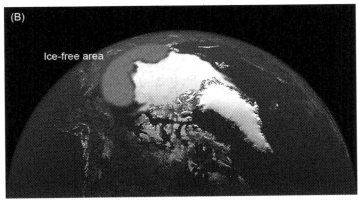

Figure 5.7 (A) Arctic sea ice, 1979.[23] (B) Arctic sea ice, 2014.

For example, Alaska's Muir glacier retreated 7 miles and saw its thickness decreased by 2625 ft., almost one-half a mile.[21] Likewise, the Columbia Glacier in Alaska has retreated 4 miles in 6 years (between 2009 and 2015) (see Fig. 5.9). This trend is evident in glaciers worldwide—in Greenland, Iceland, Peru, Argentina, Switzerland, Nepal, Italy, the United States, and elsewhere. The fabled snows of Kilimanjaro have melted by more than 80%.[22]

---

[21] Sophie Berger, "Image of the Week: 63 Years of the Muir Glacier's Retreat," EGU Blogs, October 30, 2015, https://blogs.egu.eu/divisions/cr/2015/10/30/image-of-the-week-63-years-of-the-muir-glaciers-retreat/. Three dramatic photos show extent of Muir glacier's retreat in the 63 years between 1941 and 2004.

[22] Daniel Glick, "GeoSigns: The Big Thaw," National Geographic, September 2004, p. 28, http://ngm.nationalgeographic.com/ngm/0409/feature2/fulltext.html.

[23] U.S. Environmental Protection Agency, Climate Change Indicators: Arctic Sea Ice, last updated August 2016, https://www.epa.gov/climate-indicators/climate-change-indicators-arctic-sea-ice. See NASA 2016 photographs of Arctic sea ice, September 2015 versus September 1979.

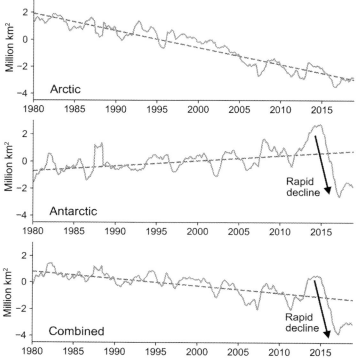
Figure 5.8 Deviation in sea ice extent (Mkm^2).[24]

The Hindu Kush-Himalayan region is home to 5500 glaciers; many are shrinking. This region, which spans eight countries from Afghanistan to Pakistan, also is the origin of 10 major rivers, including the Ganges, Indus, Yellow, Yangtze, and Mekong. Approximately 2 billion people depend on these rivers for their main source of water. A recent study of the region that took 5 years to complete warns that if the current rate of greenhouse gas emissions is not stopped, the region will warm by 5°C and lose two-thirds of its glaciers by 2100.[25] This will reduce spring river flows substantially, causing water shortages for people, animals, and crops.

---

[24] NASA, "Sea Ice Extent Sinks to Record Lows at Both Poles," March 22, 2017, https://www.nasa.gov/feature/goddard/2017/sea-ice-extent-sinks-to-record-lows-at-both-poles. Shows deviation in sea ice at both poles.

[25] Liz Boyd, "Climate Change: Himalayan Meltdown?" p. 10, *The Nation*, February 25/March 4, 2019. See also Wester, et al., *The Hindu Kush Himalaya Assessment* (New York, NY: Springer, 2019).

How do we know global warming is real?   53

Figure 5.9 The Columbia Glacier, Alaska. Left: 2009. Right: 2015.[26]

In Switzerland in 1850, glaciers covered 1735 km^2 (670 square miles) of land. Today, the area covered is about one-half of that, or 880 km^2 (340 square miles), and Swiss glaciers are melting at an unprecedented rate. The retreat of the Swiss glaciers has a macabre side effect. Swiss authorities have a map that lists everyone reported lost in the mountains since 1925 to the present day. In 2017 retreating ice revealed the bodies of a Swiss couple, missing for 75 years. A week later, the remains of a German hiker missing since 1987 were found. Like Switzerland, glaciers are also melting in the Italian Alps. Here also mummified remains are revealing themselves. There is a difference. These are World War I soldiers—victims of fighting between Italy and Austria.[27] In California, the Dana, Lyell, and Palisades glaciers, among others, have lost half or more of their mass. In Montana's Glacier National Park, the glaciers are rapidly disappearing. A century ago the park had 150 glaciers larger than 25 acres. The 25 that remain are shrinking rapidly, and park management estimates that by 2030 Glacier National Park will be without glaciers.[28]

---

[26] Michael Irving, "Before and after photos of melting glaciers capture climate change in action," *New Atlas*, April 21, 2017 (Source: James Balog and the Extreme Ice Survey), https://newatlas.com/before-after-photos-glaciers-climate-change/49143/. Used with permission from James Balog, Photographer, Earth Vision Institute.

[27] Laura Spinney, "Melting glaciers in northern Italy reveal corpses of WW1 soldiers," January 13, 2014, *The Telegraph*, http://www.telegraph.co.uk/history/world-war-one/10562017/Melting-glaciers-in-northern-Italy-reveal-corpses-of-WW1-soldiers.html.

[28] Merrit Kennedy, "Disappearing Montana Glaciers: A 'Bellwether' of Melting to Come?" May 11, 2017, *NPR*. http://www.npr.org/sections/thetwo-way/2017/05/11/527941678/disappearing-montana-glaciers-a-bellwether-of-melting-to-come.

In some cases the rapid melting of glaciers is changing the shape of the Earth's crust, according to NASA scientists. The massive weight of a glacier compresses the Earth's surface. When mass is removed, the land beneath rebounds, possibly triggering earthquakes, among other effects. The Rink Glacier in Greenland underwent rapid melting during the hot summer of 2012. The intense melting of ice produced a massive "wave" of ice and water that pushed seaward, increasing by more than 50% the amount of ice typically lost from the glacier in a year.[29]

The permafrost is melting

Permafrost is perennially frozen rock, soil, or plant material having a temperature colder than 0°C (32°F) continuously for 2 or more years. Permafrost is found primarily in Arctic regions, where it covers more than one-fourth of the exposed land in the northern hemisphere. Due to global warming, permafrost is melting all across the North, in Russia, Canada, Greenland, and Alaska. In many areas the permafrost is experiencing a pattern of thawing and freezing. As it does, the ground rises and sinks, causing roads and building foundations to buckle and crack. An even more serious consequence is that melting permafrost could release trapped methane, a potent greenhouse gas.

Extreme weather events are increasing

Global warming has begun to affect weather. As oceans become warmer, storms are becoming more frequent and severe. Hurricanes increase in strength since they derive energy and moisture from warm ocean surface waters. With severe storms, the storm surge (increased wave heights) can be 1—6 m (3.3 to almost 20 ft.). Storm surge is the greatest immediate danger to low-lying coastal areas.

---

[29] Scott Waldman, "Greenland Glacier Melt Actually Warped Earth's Crust," May 26, 2017, *Scientific American*, https://www.scientificamerican.com/article/greenland-glacier-melt-actually-warped-earths-crust/.

Weather patterns are shifting, some areas having less rainfall and suffering droughts, others experiencing higher than usual rainfall. As the Earth warms, rainfall patterns are modified, growing seasons are changed, and storms and floods are becoming more severe. As temperatures rise, more water vapor can enter the atmosphere, causing storms to become more intense and damaging. While climatologists do not say that global warming caused a particular storm, there is general agreement that heat waves, droughts, and heavier rainfall have become more frequent in the last 50 years. Global warming likely contributed to this result.[30]

Photo 5.2 Sinai desert, Egypt.

Deserts and tropics are expanding

The surface of the Earth is 71% water and 29% land. About one-third of the land area is defined as deserts, or regions having less than 254 mm (10 in.) of precipitation in the form of rain or snow per year. About one-third of the land in the United States is affected by desertification.

According to the United Nations Environment Program, the rate of desertification is speeding up. Africa is the worst affected continent, with two-thirds of its land either desert or drylands (Photo 5.2).

---

[30] April Reese, "Swell or High Water," *Scientific American* (June, 2017), 21.

The Sahara is the world's largest desert, roughly the size of China or the United States. Recent studies show that it has expanded about 10% since 1923.[31] Within the Sahara, the Sahel is a fragile southern border region with one growing season and a population estimated at 60 million. The expanding desert is encroaching on agricultural lands near the border. The loss of agriculture and livestock is one of the forces that are driving mass migrations. Deserts naturally undergo seasonal expansion and contraction, but this research indicates that the Sahara is undergoing a net expansion. Global warming is not the sole cause of increased desertification. Humans are responsible in part, through deforestation, destruction of wetlands, and poor agricultural practices such as overcropping and overgrazing.

Photo 5.3 Wild fires, California.

---

[31] Darryl Fears, "The Sahara is growing, thanks in part to climate change," *Washington Post*, March 29, 2018, reporting on report by Prof. Sumant Nigam, et al., University of Maryland, in the *Journal of Climate*, https://www.washingtonpost.com/news/energy-environment/wp/2018/03/29/the-sahara-is-growing-thanks-in-part-to-climate-change/?utm_term = .ddbcddb67994.

How do we know global warming is real? 57

Photo 5.4 Wild fires, California.

Global warming is shifting temperature zones in the Northern Hemisphere to the north. There is evidence that the climate of the tropics has moved north. The trend is such that Los Angeles will eventually have the climate of Cabo San Lucas, Mexico.[32]

There may be some beneficial changes too. For example, there may be extended growing areas in northern climes, water may become more accessible, and so on.

---

[32] Deborah Netburn, "2080 climate? Look south," p. B1, *Los Angeles Times*, February 17, 2019.

Rising temperatures are causing plant, animal, and human migration

Biologists have started reporting that certain plants are dying off in mountainous areas due to increased temperature or lack of precipitation. As a consequence of global warming, the snow line is moving higher. Likewise some animals accustomed to living at high elevations are moving to still higher elevations to find temperature zones where they can locate sources of food. Examples of this species migration can be found in Yellowstone National Park.[33] Other national parks considered at risk for species migration include Joshua Tree National Park in California, Glacier Bay in Alaska, Glacier National Park in Montana, Monteverde National Park in Costa Rica, and dozens of others.

As early as 2002, scientists were noting that the global temperature increase was having an effect on the behavior of hundreds of plant and animal species. Several types of behavioral change were identified in over 500 different species examined. The first indicator was that the density of the species at a particular location changed as they moved north or up in elevation, seeking areas within their metabolic temperature tolerance range. Field studies show that some species' ranges contracted (lower elevation limit rose), or shifted upward, in other cases. Some species seem unaffected. The second indicator was that the timing of events shifted (migration, flowering, egg laying, etc.). The third was changes in body size and behavior might occur.[34] Some species have not been affected so far.

Early warning signs of global warming: a California case history

For the last decade, the California Environmental Protection Agency's Office of Environmental Health Hazard Assessment has tracked various indicators of California's changing climate.[35] In its most recent

---

[33] Jake Abrahamson, "One Day Parts of Yellowstone May Look Like Las Vegas," *Sierra*, June 7, 2016, http://www.sierraclub.org/sierra/2016-4-july-august/americas-national-parks/what-will-climate-change-do-yellowstone.

[34] Root, et al., "Fingerprints of Global Warming on Wild Animals and Plants," Letters, *Nature*, vol. 421 (January 2003): 57–58.

[35] "Indicators of climate change in California," California Environmental Protection Agency's Office of Environmental Health Hazard Assessment, May 2018, as reported by Tony Barboza and Joe Fox, "California's changing climate is visible, from growing fires to shrinking glaciers," p. B1, *Los Angeles Times*, May 19, 2018.

report, the agency summarizes changes in key indicators. Among the most significant are the following:

- Water temperatures in the lakes and the ocean are increasing. Average lake water temperatures at Lake Tahoe have increased by nearly 1°F since 1970. During the last 4 years the warming trend has accelerated.
- Wildfires are becoming more destructive. Since 1932, 20 of the largest wildfires on record have occurred in the last two decades (Photos 5.3).
- Nights are getting hotter. Nighttime heat waves, lasting five consecutive nights or more, were once rare but have increased significantly since the 1970s.
- The ocean keeps rising, but not evenly. Local geography and tides can make a difference, even causing levels to fall in some locations. Sea levels in the San Francisco Bay have risen by 7 in. since 1900 and by 6 in. at La Jolla since 1924. The rate of rise varies up and down the coast, ranging from about 0.5 mm/year (Arena Cove) to as high as 4.7 mm/year (North Spit). The South Coast ranges from 1 mm/year (Santa Barbara) to 2.2 mm/year (San Diego). At Crescent City, the level has recently declined at 0.8 mm/year.
- California's glaciers are rapidly retreating. The Sierra Nevada glaciers have lost on average 70% of their area since 1900. The rate of decline is accelerating, with about half of the loss occurring since 1970.
- California enjoys some of the most diverse plant and animal life of any state. However, new data indicate that more than 300 species are endangered. The report indicates that Mojave Desert birds are disappearing and various types of amphibians are particularly hard hit.[36]

Overall, California is getting warmer and drier. Trees and animals are moving to high ground. More precipitation is falling as rain, and less as snow, meaning that water supplies that depend on the spring snowmelt are reduced. Drought, dead trees and dried vegetation increase the risk and size of wildfires. The cost of fighting fires has grown dramatically, as has the resulting property damage loss.

There you have it. There is a huge and growing body of evidence pointing to the fact that global warming is real. We, the authors of this book, are worried. We have lived in Southern California near the coast for decades. But we sense things are changing. It is hotter and drier. We have frequently had forest fires during the fall, but now they start in the spring, the fire season is longer, and fires are much larger. The Thomas

---

[36] Anna M. Phillips, "California species already on the brink," p. B2, *Los Angeles Times*, May 8, 2018.

Fire (one of California's largest), began in December 2017, burned 440 square miles and cost over $2 billion. We have had torrential rains and destructive mudslides. Our beaches are eroding. It is something we think about. We ask ourselves, "What are the main issues and how can we do something about them?"

CHAPTER 6

How do we know man-made CO_2 is the issue?

Some people do not believe that global warming is real. They see no evidence in their own lives that global warming is a problem. Other people believe in global warming, but think the problem is too difficult to solve. Nothing can be done in their lifetime. Some believe that it is due to natural causes, beyond the ability of humans to change. However, today there is *overwhelming* scientific evidence that global warming is due to human activities that commenced with the beginning of the Industrial Revolution.

According to the U.S. Global Climate Change Research Program report, Climate Science Special Report issued November 2018, "It is extremely likely that human activities, especially emissions of greenhouse gases, are the dominant cause of the observed warming since the mid-20th century. There is no convincing alternative explanation supported by the extent of the observational evidence."[1]

First, what about greenhouse gases? Where do they come from and where do they wind up? What happens to greenhouse gases after they enter the atmosphere?

Chapter 3, The Earth as a System, discussed the importance of greenhouse gases, mainly CO_2, to life on earth. As stated in Chapter 3, The Earth as a System, the earth's current average temperature is 15°C (59°F). Without any CO_2 in the atmosphere, the earth's surface temperature would be −18°C (0.4°F)—below freezing. Life on earth as we know it would not be possible. We need the greenhouse effect to make the earth habitable. However, excess CO_2 in the atmosphere leads to an overheated earth that also threatens life. We are not saying that global warming will end life on earth. Life will be altered. Species will disappear. We can expect a number of adverse changes in the earth's climate that are a high

---

[1] Wuebbles, et al., *Fourth National Climate Assessment (NCA4)*, U.S. Climate Science Special Report (U.S. Global Change Research Program, 2018), https://www.globalchange.gov/nca4.

Reaching Net Zero. © 2020 Elsevier Inc.
DOI: https://doi.org/10.1016/B978-0-12-823366-5.00006-3 All rights reserved. 61

risk to the way we live now. Bill McKibben's book *Eaarth* paints a vivid picture of what life on an overheated planet will be like.[2]

Let us look at greenhouse gases, and especially CO_2, in more detail since CO_2 is the greatest contributor to global warming.

Where do man-made greenhouse gases come from?

In 2018, global emissions of greenhouse gases totaled 55 $GmtCO_{2eq}$, an increase of about 2.7% over 2017, the largest increase in seven years. About 75% of these emissions were due to fossil fuel use to generate power and heat for transportation, industry, commercial, and residential uses. Twenty-five percent came from agriculture and land-use changes.

Fossil fuels produce 85 percent of the world's energy. The remainder is produced by nuclear power or hydropower. Renewable energy, mainly from solar panels and wind turbines and a few other sources, provides a small portion of the world's energy today. However, renewables to produce electricity are the fastest growing source of energy in the world. The cost of electricity from solar panels and wind turbines has decreased significantly in the last 20 years and continues to improve. This makes electricity from renewables the cheapest source of energy compared to fossil fuels and nuclear power in an increasing number of locations.

For energy production, the use of coal and natural gas to produce electricity is the single largest use of fossil fuels but is being replaced by renewables as the cost of renewables continues to decline.

The next largest use of fossil fuels is for transportation. Most of this is in the form of liquid fuels, gasoline, diesel fuel, aviation fuel, and liquefied natural gas. These are produced by refining oil.

Industry is the next largest energy user. Some fossil fuels, mainly oil and natural gas, are used as feedstock to produce plastics, fertilizers, and other chemicals. Most are used to produce electricity and heat to power industrial processes. Making steel and cement are the largest industrial energy users. Aluminum production is energy-intensive but it is primarily produced with hydroelectricity.

---

[2] Bill McKibben, *Eaarth: Making a life on a tough new planet,* (New York, NY: Times Books, 2010).

Residential and commercial uses include lighting, heating and cooling buildings, hot water heating, and cooking. Other miscellaneous uses account for the balance.

Any plan to stop global warming has to take into consideration agriculture and land use changes that account for about 25 percent of global greenhouse gas emissions. These emissions are not reduced much if at all by actions taken to cut emissions from the use of fossil fuels. If emissions from post-production of food, and other activities outside the farm are included, over 30 percent of emissions could be due to this sector. Post-production includes food processing, transportation, storage, and other activities related to getting food from farms to markets.

The IPCC refers to these emissions as coming from agriculture, forestry, and other land use (AFOLU). Because agriculture, forestry, and land use are so interrelated, their emissions are often dealt with as a single source. According to the IPCC Special Report on Climate Change and Land, 12.0 Gmt/year were due to AFOLU. Emissions due to agriculture were 6.2 Gmt/year and emissions due to land use changes were 5.8 Gmt/year.[3]

Greenhouse gas emissions from agriculture have been increasing steadily at about 1.0 percent per year, roughly in line with global population growth. Emissions from deforestation and other land use changes have been steady for a long time.

AFOLU emissions account for about 44 percent of total methane emissions, about 81 percent of nitrous oxide emissions, and about 11 percent of CO_2 emissions. Most methane emissions are due to raising cattle and for rice farming. Most nitrous oxide emissions are associated with the use of nitrogen fertilizers. CO_2 emissions are associated with fossil fuel used in farming and forest clearing and indirectly in manufacturing fertilizer and pesticides.

As shown in Figure 6.3, about 30% of total CO_2 emissions, about 12 Gmt/year, are absorbed by plants and soils. Rain forests in Brazil and Indonesia are large carbon sinks. Land use changes, mainly deforestation, reduce the amount of CO_2 that can be absorbed by plants and soils. Other land use changes add to emissions, for example, by burning crop residues and felled trees, slash and burn agriculture, and forest fires.

---

[3] IPCC Special Report on Climate Change and Land, (A report addressing greenhouse gas (GHG) fluxes in land-based ecosystems, land use and sustainable land management in relation to climate change), August 8, 2019, https://www.ipcc.ch/srccl/

About 80 percent of forest clearing is to create more land for agriculture. Most of this land is for grazing cattle and for raising feed for cattle. This is the leading cause of deforestation. It is estimated that meat production alone is responsible for 14 percent of global greenhouse gas emissions, mostly as methane. Logging and mining also lead to deforestation but are much less a factor than clearing land for agriculture. In Southeast Asia, forests are cleared to create land to grow palms for oil which is used for cooking, to produce biodiesel fuel used in Europe, and for other commercial purposes.

Between 1990 and 2016 the world lost an estimated 1.3 million square kilometers of forests, an area larger than South Africa. Unfortunately, the annual destruction of forests is increasing and emissions from land use changes could increase in the future. In 2018, an estimated 120,000 square kilometers, an area four times the size of Belgium, was lost to deforestation. Most of the loss was in Brazil, The Democratic Republic of Congo, and Indonesia. These and other countries have few incentives and limited means to implement actions to improve farming and reduce deforestation. There is a need for the use of carbon credits or other means to provide financial incentives to reduce emissions and also pay for the cost of implementation, monitoring, and enforcement programs. Part of revenues from carbon fees should be allocated for this purpose.

Deforestation, other land clearing, and some agricultural practices remove trees and other plant life and thereby contribute to global warming by removing a CO_2 "sink," meaning something that absorbs CO_2. If the earth's trees and plants sequester (absorb) less CO_2, then the difference has to be absorbed by the oceans and the remainder stays in the atmosphere. The amount staying in the atmosphere increases CO_2 concentration and increases the greenhouse effect. Forest fires have the same effect as deforestation with the added problem of releasing CO_2 and other pollutants into the atmosphere, the same as burning fossil fuels. The one difference is that burned forests will slowly sequester (absorb) carbon over several decades as they grow back (Photo 6.1).

CO_2, primarily from the combustion of fossil fuels, is the largest greenhouse gas by volume. Rather than list all the greenhouse gases individually, scientists refer to CO_{2eq}, which includes the other greenhouse gases converted to a CO_2 equivalent. CO_{2eq} is roughly equal to 1.32 times CO_2 alone. Throughout this book, when we refer to CO_{2eq}, it is CO_2 plus other greenhouse gases. See Table 6.1 for a summary of principal greenhouse gas sources. Figs. 6.1 and 6.2 show components and origins of greenhouse gas.

How do we know man-made CO_2 is the issue?

Photo 6.1 Clearing tropics.

Table 6.1 Greenhouse gas sources.

Greenhouse gas	Percent of total	Source
Carbon dioxide (CO_2)	76	Fossil fuel use (coal, oil, natural gas)
		Deforestation and other land-use changes
Methane (CH_4)	16	Cattle raising
		Waste decomposition
		Natural gas leakage
		Rice farming
		Swamp gas emissions
Nitrous oxide (N_2O)	6	Fertilizer use and other agricultural activities
		Fossil fuel use
Fluorinated gases (F-gases)	2	Semiconductor and aluminum manufacturing
		Other industrial processes

What happens to CO_2 emissions?

CO_2 does not go away. The removal of excess CO_2 from the atmosphere by natural processes is very slow. After 100 years, 40% or 50% of the CO_2 is still in the atmosphere reflecting solar energy back to earth. Due to latency, it will take a long time for the full effects of a higher CO_2 concentration to show up as climate changes.

Not all CO_2 is added to the atmosphere as a greenhouse gas. About 30% is consumed by growth of trees and plants. Growing trees and plants

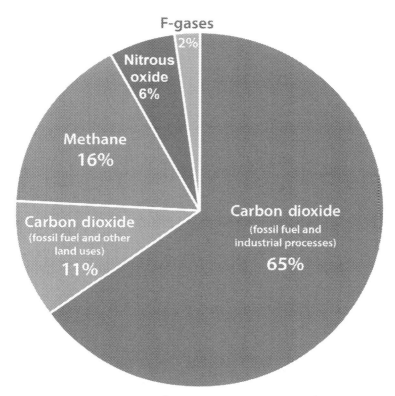

Figure 6.1 CO_{2eq} gas components.[4]

absorb CO_2 from the atmosphere by way of photosynthesis to provide plants the energy they need to grow. Plants absorb CO_2, use the carbon for plant growth, and release O_2. Absorbed carbon is sequestered until the trees and plants are destroyed by fire or die and decompose.

Another 30% is absorbed by the oceans. The ocean takes up CO_2 through diffusion and photosynthesis by plant-like organisms (phytoplankton) and is necessary for shellfish growth. Some CO_2 forms carbonic acid, which raises the acidity of the oceans. Increasing acidity *inhibits* shell growth of marine animals and is destroying tropical reefs, which provide essential habitat for marine life.

---

[4] Global greenhouse gas emissions data (by type of gas), EPA (Source: *IPCC 2014*), https://www.epa.gov/ghgemissions/global-greenhouse-gas-emissions-data.

How do we know man-made CO_2 is the issue?

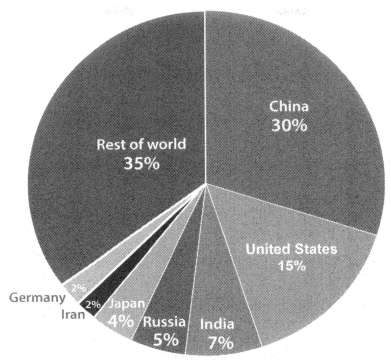

Figure 6.2 Global CO_2 emissions by world region, 2014.[5]

The rest is absorbed by slower processes that take as long as decades to thousands of years, including chemical weathering and rock formation (see Fig. 6.3).

In addition to absorbing CO_2, the oceans also absorb most of the *heat* produced by global warming. It is estimated that over 90% of the additional heat from man-made CO_2 emissions is stored in the oceans.[6] Warmer seawater alters fish migration and imperils giant kelp forests along California's coast. Some of this heat goes into melting sea ice, ice caps, and glaciers. Some is absorbed by heating the landmass. Only a small amount of the excess heat—less than 10%—goes directly into warming

---

[5] Global CO_2 emissions by world region (2014), EPA (Source: Boden, T.A., Marland, G., and Andres, R.J. (2017). *National CO_2 Emissions from Fossil-Fuel Burning, Cement Manufacture, and Gas Flaring: 1751–2014*, Carbon Dioxide Information Analysis Center, Oak Ridge National Laboratory, U.S. Department of Energy. doi:10.3334/CDIAC/00001_V2017. https://www.epa.gov/ghgemissions/global-greenhouse-gas-emissions-data.
[6] https://www.oceanscientists.org/index.php/topics/ocean-warming. Data from IPCC 5th assessment report.

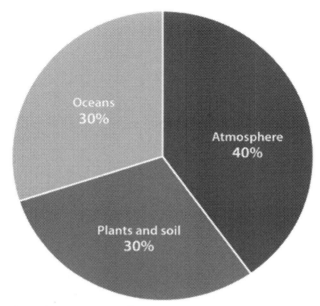

Figure 6.3 Where do emissions go?[7]

the atmosphere, yet that 10% is extremely significant as it relates to temperatures on earth.[8]

The significance of carbon-14

(Textbox 6.1)

> **TEXTBOX 6.1**
> What is an isotope? The atomic weight of an element is determined by the number of neutrons and protons in the nucleus. "Normal" carbon has six protons and six neutrons, is denoted as C-12, and is stable. Carbon-14 has six protons and eight neutrons, making it unstable and radioactive.

---

[7] National Center for Atmospheric Research, University Cooperation for Atmospheric Research, Center for Science Education. Where does carbon dioxide go? https://scied.ucar.edu/imagecontent/where-does-carbon-dioxide-go.

[8] LuAnn Dahlman and Rebecca Lindsey, "Climate change: ocean heat content," *NOAA, climate.gov*, August 1, 2018, https://www.climate.gov/news-features/understanding-climate/climate-change-ocean-heat-content.

The concentration in the atmosphere of carbon–14 (C–14), the radioactive isotope of carbon, is decreasing. As strange as it may seem, this is definitive proof that human activity is causing global warming. The explanation is complicated.

A bit of history first: in 1949 Willard Libby invented carbon dating, for which he won a Nobel Prize in 1960. Carbon-14, which is radioactive with a 5730-year half-life, does not exist in nature. It is formed in the atmosphere when cosmic rays interact with nitrogen (80% of air), converting nitrogen-14 (N–14) to C–14. Most CO_2 molecules consist of *carbon-12* (C–12) plus two atoms of oxygen. About one out of every *trillion* CO_2 molecules will include C–14, rather than the common (and nonradioactive) C–12 isotope. Atmospheric CO_2 is absorbed by plants through photosynthesis and by the ocean, and becomes carbohydrates in animals that consume plants.

Here is Libby's discovery: when a plant dies, or piece of firewood is cut from a tree, it no longer absorbs CO_2 and therefore no longer takes up any C–14. The C–14 it contains begins to undergo radioactive decay and 5730 years later half of it will be gone. By sensitive measurements, Libby could calculate how much C–14 remained (compared to the amount originally absorbed, which was known from atmospheric data) and he could thus determine the age of the sample.

In 1955, a scientist named Hans Suess discovered that the atmospheric concentration of C–14 had decreased by 2.5% from the value in 1890.[9] Since the earlier date followed shortly after the beginning of the Industrial Revolution in the United States, he postulated that what was happening was the combustion of fossil fuels (coal, oil, and natural gas, but not firewood) was diluting the CO_2 in the atmosphere, so the ratio of C–12 to C–14 was *increasing*.

Then something happened.

Fast forward to the 1960s and atmospheric tests of nuclear weapons. Radiation from nuclear explosions produced C–14 that was released to the atmosphere and suddenly the concentration was about 1½ times *more* than the historic value. This caused renewed interest in monitoring C–14 levels. Between 1965 and 1995, the C–14 concentration gradually decreased, due to dilution by CO_2 from fossil fuel emissions (pure C–12, with no C–14).

---

[9] Hans E. Suess, "Letters" (On the measurement of C-14 in the atmosphere), *Science*, 122, no. 316 (September 1955): 415–416.

Oil, natural gas, and coal are called *fossil* fuels, because they are derived from the decayed remains of ancient plants and animals that died millions of years ago. They were formed so long ago that any C-14 has completely decayed. Fossil fuels only contain C-12. Suess saw the decrease in atmospheric C-14 as proof-positive that human activities were increasing the amount of atmospheric CO_2, thereby diluting the amount of C-14 compared to its historical equilibrium value. Declining C-14 concentration is a certain indicator that human activity is increasing greenhouse gases.[10]

Historic emissions since the Industrial Revolution

Today, there are about 3200 Gmt of CO_2 in the atmosphere. See Appendix 2. CO_2 persists in the atmosphere for hundreds of years. Only about $10 - 11$ Gmt (gigametric tons) of CO_2 are removed from the atmosphere each year by natural processes. Any man-made or natural CO_2 emissions that exceed this amount will cause the amount of CO_2 in the atmosphere to increase.

Global manmade CO_2 emission in 2018 was 37.1 Gmt, about three times greater than CO_2 leaving the atmosphere. To prevent any further CO_2 buildup in the atmosphere, fossil fuel use would have to be eliminated. We would have to get to net zero according to the IPCC, because any man-made emissions would have to be matched by sinks that absorb CO_2, such as growing forests and other plant life. Artificial methods to remove CO_2 from the atmosphere are being discussed but so far are not technically feasible or economical on the scale that would be required.

To understand the magnitude of this problem, consider this: *if every country in the world stopped consuming oil tomorrow,* man-made emissions would only decline by about one-third, or 13 Gmt of CO_2/year. Now imagine how hard it will be—and how long it will take—just to stop using oil.

The effects on the climate from past excess CO_{2eq} emissions will persist for many years until natural absorption of excess CO_2 restores concentrations to preindustrial levels—about 280 ppm. The earth's temperature is already $1.0°C$ above preindustrial levels, and it will take a long time for the earth's temperature to return to its previous average value.

---

[10] Heather D. Graven, "Impact of fossil fuel emissions on atmospheric radiocarbon and various applications of radiocarbon over this century," *PNAS*, 112, no. 31 (August 2015): 9542–9545.

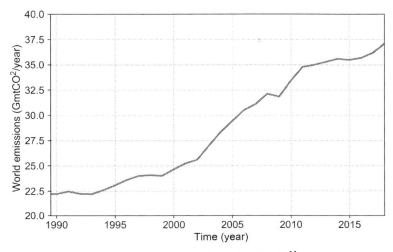

Figure 6.4 Global carbon dioxide emissions from fossil fuels.[11]

Three countries and the European Union (EU) produce two-thirds of global CO_2 emissions. They are, in order of importance, China, United States, the EU, and India (see Fig. 4.6). In less than a decade China's CO_2 emissions have doubled! It has now surpassed the United States as the world's leading atmospheric contributor of CO_2.[12]

Consider now Fig. 6.4. This shows that global CO_2 emissions peaked in 2014, and then dipped briefly in 2015 and 2016.[13] This may have been due to the increased use of natural gas displacing coal and the increased use of renewable energy. In 2017, greenhouse gases increased 1.6% and in 2018, rose to 37.1 Gmt of CO_2, creating a new record. Fig. 6.4 data are for CO_2 from combustion of solid, liquid, and gas fossil fuels, as well as cement production and gas flaring. Total greenhouse gases (CO_{2eq}, including methane and other gases) reached 55 Gmt in 2018.[14]

---

[11] T.A. Boden, et al. (2010), "Global, Regional, and National Fossil-Fuel CO_2 Emissions," Carbon Dioxide Information Analysis Center, Oak Ridge National Laboratory, Oak Ridge, TN. doi:10.3334/CDIAC/00001_v2010.

[12] Glen Peters, "Have Chinese CO_2 emissions really peaked?" *Climate Change News* (March 31, 2017), http://www.climatechangenews.com/2017/03/31/chinese-co2-emissions-really-peaked/.

[13] Global CO_2 emissions, 1980−2016, IEA, https://www.iea.org/newsroom/energysnapshots/global-carbon-dioxide-emissions-1980−2016.html.

[14] "Trends in global CO_2 and total greenhouse gas emissions," *Netherlands Environmental Assessment Agency*, https://www.pbl.nl/en/publications/trends-in-global-co2-and-total-greenhouse-gas-emissions-2018-report. Also see https://www.ipcc.ch/site/assets/uploads/2018/12/UNEP-1.pdf and Climate Interactive for 2018 data.

Increases in atmospheric CO_2 correlates with fossil fuel use

As Fig. 6.4 shows, CO_2 from fossil fuels began increasing after 1900, due largely to the rise of the oil industry in the United States. The post-Industrial Revolution society has continuously increased the combustion of fossil fuels—coal, natural gas, and petroleum—and their derivatives, gasoline, diesel fuel, heating oil, and compressed liquefied natural gas. Combustion of fossil fuels basically involves combining oxygen and carbon to release the energy found in the fuel molecules, which contain carbon, hydrogen, and other substances. In this process, carbon dioxide is formed, along with some carbon monoxide. In the atmosphere, carbon monoxide is converted to carbon dioxide.

Burning fossil fuels produces varying amounts of CO_2, depending on the fuel type and purity. Representative yields of CO_2 to generate 1 GJ (0.948 MBtu) of energy by burning fossil fuel are:
- Coal—90 kg of CO_2
- Diesel—73 kg CO_2
- Natural gas—50 kg of CO_2

By calculating how much fossil fuel is used each year and comparing the amount of CO_2 produced with the observed increase of CO_2 in the atmosphere, scientists conclude that the increase is due to fossil fuel combustion. This leads us to the inescapable conclusion that the increase in greenhouse gas production is due to human activity. In other words, "We have met the enemy and they are us." (Apologies to cartoonist Walt Kelly and his strip "Pogo.")

What about natural sources of CO_2? Forest fires contribute to the CO_2 entering the atmosphere, but the amount is small compared to fossil fuel use. Likewise, volcanoes emit CO_2, but only one-hundredth or less when compared on an annual basis to the amount due to burning fossil fuels.[15]

---

[15] Shahir Masri, *Beyond Debate: Answers to 50 Misconceptions on Climate Change* (Newport Beach, CA: Dockside Sailing Press, 2018), 31–33.

What is your carbon footprint?

Each of us contributes to global warming every day through our daily activities. We probably have control of about one-fourth of our per capita energy use in the United States, that being our personal home use of energy for transportation, lighting, space conditioning, and cooking, for example. The rest is "indirect," meaning not literally under an individual's direct control, but can be influenced by lifestyle choices. Indirect energy is due to our share of energy used for food production, purchased manufactured goods, heating, cooling, and lighting of commercial and government buildings and public spaces, commercial freight, water and sewage treatment, and so on. How much do we individually contribute to global warming via our discretionary use of fossil fuels? Lifestyle is an important determinant of human CO_2 emissions and is a way we can influence emissions. For example, a hunter–gatherer living in the Amazon forest today has an energy input of 5−7 GJ/year per person for survival, plus net CO_2 emissions (wood burning) of 0.24 mtCO_2/year per person. Contrast this to the average American, who used 300 GJ/person per year in 2016 and emitted 20 mtCO_2/year per person.[16]

The U.S. Environmental Protection Agency has an on-line program that you can use to calculate your carbon footprint: https://www3.epa. gov/carbon-footprint-calculator/.

These differences highlight the dilemma we face. There are poor and underdeveloped countries that realistically need to dramatically increase their energy use, mainly electricity for lighting, gasoline for transportation, and natural gas for cooking and heating. Today, many of these countries principally depend on firewood, dung, and wood-produced charcoal for their meager energy diet. The use of wood and charcoal is destroying remaining forests and increasing atmospheric CO_2 levels. In addition, burning wood and charcoal in homes causes indoor air pollution that leads to health problems. In setting goals for reducing CO_2 production, we need to account for the need to *increase* energy use in many countries, including some large countries such as India. A better approach would be to bypass increased fossil fuel use and go directly to more renewable energy.

---

[16] Sources: For U.S. per capita energy use, https://www.eia.gov/tools/faqs/faq.php?id = 85&t = 1; for average CO_2 emissions, see: https://www.sciencedaily.com/releases/2008/04/080428120658. htm.

CHAPTER 7

What are the effects of global warming?

It has been established that increases in greenhouse gas concentrations in the atmosphere lead to increases in the earth's temperature, which has an increasingly adverse effect on the earth's climate and population.

The Intergovernmental Panel on Climate Change (IPCC) recommends that actions be taken to limit the earth's temperature increase to no more than 2.0°C, and preferably to no more than 1.5°C above preindustrial levels. However, in its October *2018 Special Report: Warming of 1.5°C*, the IPCC states with a high degree of confidence that global warming is *likely* to reach 1.5°C between 2030 and 2052 if it continues to increase at the current rate. Limiting global warming to 1.5°C is probably no longer achievable.[1]

The 2.0°C limit is significant. It is not a sharp threshold beyond which something terrible will happen. It is considered to be the upper limit of present-day natural variability, especially in the tropics. Exceeding 2.0°C will cause greater damage from global warming, such as more severe heat waves, lower crop yields, and an almost complete bleaching of the coral in the ocean. By greatly exceeding natural variability, it increases the uncertainties in trying to predict the effects of global warming.

In thinking about the effects of global warming, readers living in developed countries in Europe and North America must keep in mind that much of the world is at far greater risk from global warming. According to the World Bank, about 45% of the world's population lives on $5.50 per day or less.[2] They have problems with housing, sanitation, access to clean water and medical care and are generally vulnerable to adverse climate changes. About two billion people are subsistence farmers. They survive on small landholdings producing most of what they need to

---

[1] Op. cit. V. Masson-Delmotte, et al., *IPCCSR5.1. An IPCC Special Report on the Impacts of Global Warming of 1.5°C above Pre-industrial Levels and Related Global Greenhouse Gas Emission Pathways, in the Context of Strengthening the Global Response to the Threat of Climate Change, Sustainable Development, and Efforts to Eradicate Poverty* (Geneva, Switzerland: World Meteorological Organization, 2018), https://www.ipcc.ch/sr15/.

[2] See note at bottom of Table 4.1.

survive. Another billion people depend on fishing as their primary source of animal protein.

Latency—how long before effects show up?

For most of the potential risks, there is a delay (latency) between cause and effect. Chapter 1, Introduction, cited the example of lung cancer deaths and cigarette smoking. In many cases, the causes began having an effect years ago, and the results are beginning to show up now. However, the ultimate impacts of global warming will be unknown for some time. We cannot be absolutely certain that any effort to reduce greenhouse gas emissions will be effective in a realistic time frame. Even if (hypothetically) we were to completely stop all releases of CO_2 to the atmosphere tomorrow, global temperatures would remain high for decades. As much as a century or more might elapse before the CO_2 concentration in the atmosphere began to decline enough to lower global temperatures.

Climate change versus weather

As stated in the *Preface*, it is important not to confuse climate change with weather. Climate is changing due to greenhouse gases raising the earth's temperature, a process known as global warming. This book focuses on the effects of global warming.

In most locations on the earth, the weather can exhibit wide variations and can change suddenly and violently such as during a major storm. Extreme weather events occur infrequently and, for example, are referred to as 100-year storms or floods. Because of the wide and unpredictable nature of weather extremes, it is difficult to link weather changes directly to global warming. We know that ocean temperatures influence hurricanes and other storms. However, it is not possible today to prove that a specific storm was directly due to global warming. We must look at changes in frequency and patterns.

As an example, *Hurricane Harvey* dumped a record 27 trillion gallons of water on Texas in August 2017.[3] It was a Category 4 storm that caused an estimated $125 million in damages and killed about 90 people. *Harvey* dropped 2 ft. of water on Houston in the first 24 hours. *Harvey* lingered over Houston for 4 days. At Nederland, Texas, *Harvey* set a record for a single storm in the continental United States that created an unprecedented 1000-year flood event when 64.5 in. of rain fell over 5 days. Nothing of that size has happened within modern recorded history. Flooding covered an area of southeast Texas the size of Slovenia or the state of New Jersey.

It was later determined that this storm was an estimated 38% stronger due to warming of the water in the Gulf of Mexico and warmer air temperatures attributed to global warming. Warmer water releases more moisture into the atmosphere. The warmer air can absorb more moisture that is released when the storm hits the coast where the moisture is deposited as rain. It was also determined that the amount of moisture picked up by the hurricane equaled the amount of rainfall on Texas.

We can see the impact of global warming on our climate, the average of our weather over time. Some typical climate measurements are atmospheric and ocean temperatures, melting glaciers and sea ice, and other parameters.

While a great deal is known about weather and climate today, we may be in for some surprises. Some effects may be nonlinear or involve positive feedback (e.g., warming increases melting of polar ice, which changes seawater temperature, thereby increasing melting, etc.). For example, as more ice melts, the albedo (reflectivity) of the surface changes, thus facilitating more energy absorption rather than reflection, and increases warming. Some effects could turn out to be irreversible. In the Arctic, microorganisms break down organic matter, releasing methane, a potent greenhouse gas. Millions of tons of carbon are trapped in the ice layer under permafrost in the form of methane clathrate (Textbox 7.1).

Should this protective layer melt, methane could be released to the atmosphere, thereby causing more warming that could release still more gas, and so on. One cubic meter (m^3) of methane clathrate releases $160 \ m^3$ of methane gas. This subject is of such importance that Chapter 9, What Would It Take to Reach Net Zero?, provides more detail.

---

[3] Henry Fountain, "Scientists Link Hurricane Harvey's Record Rainfall to Climate Change," *The New York Times*, December 13, 2017.

> **TEXTBOX 7.1**
> Methane is the main component of natural gas. Methane clathrate is methane trapped within a crystal structure of water, forming an ice-like solid.

As we point out in Chapter 5, How Do We Know Global Warming Is Real?, there is a growing body of solid scientific evidence that global warming is a real concern and should not be ignored. For that reason, it is important to examine what unfavorable consequences are occurring already and what future dangers exist for humanity. Here is what we perceive to be the most likely risks.

Earth's temperature will continue to rise

As noted in Chapter 5, How Do We Know Global Warming Is Real?, earth's average temperature is rising. It will continue to rise until greenhouse gas emissions cease and the atmospheric concentration of CO_2 falls below 300 ppm. This will take a century or more.

Rising temperatures have a major impact on health and human productivity. Heat waves affect people in a number of ways, including heat stress, heatstroke, and dehydration. The impact is especially severe for children, pregnant women, and the elderly. Labor productivity is reduced as well, as temperatures increase, especially for those working outdoors and manual laborers. Lower productivity affects incomes. This effect is especially acute for communities that rely on subsistence farming that requires hard, manual labor.

Canada provides a good example. Canadian government studies indicate that Canada is warming at twice the rate of the rest of the world. Its average land temperature has increased 1.7°C (3°F) since 1948, when it began keeping records. Temperatures have shown even greater increases in the colder northern part of the country. This is due in part to melting glaciers and less snowfall and raises the risk of droughts and wildfires.[4]

---

[4] Health and Science News, "Canada Getting Hotter Faster," *The Week*, April 19, 2019, p. 22.

Air pollution will increase

Air pollution will increase, especially in rapidly growing megacities in Asia. As countries such as China and India use more coal to meet their growing demand for electricity, air quality will decline, even as other countries expand the use of nonpolluting renewable energy sources. Air pollution from motor vehicles will also increase as rising living standards cause more people to own motor vehicles. Air pollution knows no national boundaries. Studies show that pollutants drift across the ocean and circulate in the northern hemisphere. In one case, pollution released in East Asia was detected in central Oregon 8 days later.

A 2015 study from the nonprofit organization *Berkeley Earth* estimated that 1.6 million people in China die each year from heart, lung, and stroke problems because of polluted air.[5] India has now surpassed China as having the most polluted cities. Delhi is exposed to toxic emissions from traffic, home trash burning and cooking smoke, rubbish burning, and agriculture burning. Gary Fuller, air pollution specialist at King's College London, calls this "the invisible killer." He says, "We need to stop using the air that we breathe as a waste disposal route." Otherwise, the more than four million people who die prematurely each year globally will continue to increase.[6]

Sea levels rise causing flooding

The U.S. population living within coastal counties from Maine to Florida, along the Gulf States, and along the Pacific coast, is approximately 123 million people. An estimated 13 million people will be in the path of flooding by 2100, if not sooner. For the United States, 4.9 million people would be affected by a sea level rise of 0.9 m (2.95 ft.). This would increase to 13.1 million for a rise of 1.8 m (5.9 ft.). The U.S. National

---

[5] Associated Press, "Air pollution in China is Killing 4,000 People Every Day, a New Study Finds." *The Guardian*, August 8, 2013.

[6] Gary Fuller, "The Invisible Killer," *Geographical*, 91, no. 1 (January 2019): 13.

Oceanic and Atmospheric Administration (NOAA) reports that 0.2 m (8 in.) to 2.0 m (6.6 ft.) by 2100 is possible, depending on the speed at which polar ice melts.[7]

Florida is an area that is particularly at risk. Sea level rise is threatening a huge coastal population in South Florida. More than half the population of over 100 Florida cities and towns live on land that is less than 4 ft. above high tide. There is a high probability that their homes will experience flooding by 2050. A combination of rising seas and a storm surge could cause flooding as early as 2030.[8]

The southeastern Virginia coast is among the most threatened U.S. areas for damage by sea level rise. The Port of Virginia—fifth largest container ship entry point into the United States—recently began a $375 million renovation. The plans include raising the port's electric power stations several feet and moving critical data servers inland to higher ground.[9] Norfolk, Virginia is a low-lying city on the Chesapeake Bay that is vulnerable to tidal flooding. Based on a more recent assessment, coastal communities can expect waters to rise as much as 3.5 m (11.5 ft.) by 2100.[10] Norfolk is an important base for the U.S. Navy, and as a sign of concern, the Navy is already taking steps to protect critical facilities from the impact of rising seas.

On a global scale, similar concerns prevail. More than 100 million people worldwide live within 3 ft. of mean sea level. Especially vulnerable are Shanghai, Bangkok, and Bangladesh[11] (Photo 7.1).

Rising seas are overwhelming islands in the South Pacific. Five islands in the Solomon Islands group have disappeared, with six more severely eroded.[12] The island republic of Tuvalu has been making plans to evacuate residents from low-lying areas. Kiribati is likely to become uninhabitable due to saltwater intrusion even before it is overrun by

---

[7] Marianne Lavelle, "Americans in Danger from Rising Seas could Triple," *National Geographic*, March 14, 2016, http://news.nationalgeographic.com/2016/03/160314-rising-seas-US-climate-flooding-florida/.

[8] Ben Strauss, "Florida and the Rising Sea," Climate Central-Surging Seas, accessed February 8, 2020, http://sealevel.climatecentral.org/news/floria-and-the-rising-sea.

[9] Erica E. Phillips, "Seaports Add Protections as Ocean Levels Rise," *Wall Street Journal*, February 11, 2019.

[10] April Reese, "Swell or High Water," *Scientific American, 316, no. 6* (June 2017): 21.

[11] Glick, *National Geographic*, p. 28.

[12] Simon Albert, et al., "Sea Level Rise Swallows 5 Whole Pacific Islands," *The Conversation, Scientific American*, May 9, 2016.

What are the effects of global warming? 81

Photo 7.1 Flooding in Newport Beach, California.

the sea.[13] Studies conducted in the Pacific Ocean found sea level rise at the rate of 3—5 mm/year, compared to 1.5—3 mm/year in other areas. The sea level rise in the Solomon Islands has been 7—10 mm/year since 1993. These higher rates are what can be expected across much of the Pacific in the second half of this century. The *rate* of sea level rise has been increasing and is now about three times greater than a decade ago.

Today, much of the Carteret Island group in the South Pacific has been overtaken by rising seas. Most of the population is relocating to higher ground on Bougainville Island. According to the Australian National Tide Facility, the annual sea level rise has been 8.2 mm/year for every year they have monitored. The Carteret Islanders are said to be the world's first *climate change refugees*.[14]

That claim may not be correct. Today 15,000 immigrants from the Marshall Islands live in Arkansas, about 20% of the population. Their former home is a group of low-lying atolls spread over 750,000 square miles of ocean. They have their own churches, a radio station, and even a

---

[13] Union of Concerned Scientists, Climate Hot Map, "Republic of Kiribati," accessed February 8, 2020, http://www.climatehotmap.org/global-warming-locations/republic-of-kiribati.html.

[14] Brian Merchant, "First Official Climate Change Refugees Evacuate Their Island Homes for Good," *Earth First Newswire*, April 6, 2014, http://earthfirstjournal.org/newswire/2014/04/06/first-official-climate-change-refugees-evacuate-their-island-homes-for-good.

> **TEXTBOX 7.2**
>
> Timing is everything....
>
> On August 24, 1942, Airman Delmar D. Wiley was the only survivor when his Navy plane was shot down by the Japanese in the Solomon Islands. Miraculously, Wiley managed to scramble into a small rubber raft that floated free from the sinking aircraft. With wounds in his thigh and ankle, no food, and three canteens of water, Wiley drifted 400 miles in 15 days before landing on a speck of land in the Carteret Islands, 53 miles northeast of Bougainville, where natives nursed him back to health. Otherwise, he figured he would have lived only one more day. If Wiley crashed today under similar circumstances, he would drift on to his death—the island where he landed in 1942 is under water.

consulate in Springdale, Arkansas. The atolls they are leaving are only a few feet above mean sea level, and the highest point anywhere is a hill about 9.7 m (32 ft.) high. As writer Kenneth Brower states, they are "surfing in on the first wave of what will be a global tsunami of climate refugees."[15]

The Center for Climate Integrity recently released a detailed cost estimate for constructing seawalls in the United States during the next 20 years. The bill for the 10 states along the east and west coasts and the Gulf of Mexico is $416 billion. Florida is at greatest risk, facing an expense of $76 billion. California comes in at $22 billion[16] (Textbox 7.2).

Living in Newport Beach California, as we (the authors) do, we are familiar with sea rise and coastal flooding. It happens regularly on the Balboa Peninsula when a storm surge coincides with a high tide. Balboa Island's aging seawalls have been overrun with increasing frequency in recent years. Winter storms have taken a toll on the southern California coast from Santa Barbara to San Diego. The Ventura public pier suffered wave damage and is closed. Beachfront cliffs in Del Mar and San Jua Capistrano are eroding, threatening expensive beachfront structures. In Newport Beach, city government is reluctant to confront the problem because of the potential costs involved, but recently undertook some minimal repairs on Balboa Island's 80—90-year-old seawalls. At best, this will buy a few years.

---

[15] Kenneth Brower, "The Atolls of Kansas: Climate Refugees from the Marshall Islands Find a New Home in Springdale," *Sierra* (January/February 2019): 20—26.

[16] See Center for Climate Integrity, Climate Costs in 2040, accessed February 8, 2020, http://climatecosts2040.org/.

This brings up an important point. Market forces are much better at recognizing risk than politicians. Prices are falling for expensive oceanfront homes as the seas rise and the frequency of storms increases. Residences that are exposed to the potential of rising water sell at a 7% discount compared to waterfront homes that are better protected.[17]

Beyond housing, a huge amount of coastal infrastructure—power systems, water and sewer, highways and rail, and docks—is threatened by rising seas.

Water-borne infectious diseases are spread by flooding. Flooding caused by storms damages coastal infrastructure and water and sewage systems. As a result, water supplies can become contaminated.

Refer to Jeff Goodell's book (*The Water Will Come: Rising Seas, Sinking Cities, and the Remaking of the Civilized World*) for a detailed discussion of rising sea levels.[18]

Oceans become more acidic

The CO_2 levels in the ocean are rising at the same rate as the atmospheric levels, except that deep water is accumulating CO_2 at a faster rate. When CO_2 dissolves in saltwater, carbonic acid is produced. This increased acidification has the potential to kill coral, a living organism and a vital part of the food chain that supports the world's fisheries. Reefs are dying; half of the Great Barrier Reef in Australia has been bleached to death.[19]

Increased CO_2 levels have caused lower aragonite (a form of calcium carbonate) saturation levels in the oceans around the world. This makes it more difficult for marine organisms to build shells and skeletons.[20] Increased CO_2 and excess nutrients from fertilizer runoff and overuse

---

[17] As reported by Ed Leefeldt in *CBSNews.com*. See also David Z. Morris, "Climate Change is Already Depressing the Price of Flood-Prone Real Estate," *Fortune Magazine*, April 21, 2018.

[18] Jeff Goodell, Jeff (2017) *The Water Will Come: Rising Seas, Sinking Cities, and the Remaking of the Civilized World*, New York, NY: Little, Brown and Company (Hachette Book Group, 2017).

[19] Lauren E. James, "Half the Great Barrier Reef is Dead," *National Geographic Magazine*, August 2018, https://www.nationalgeographic.com/magazine/2018/08/explore-atlas-great-barrier-reef-coral-bleaching-map-climate-change/.

[20] U.S. Environmental Protection Agency, Climate Change Indicators: Ocean Acidity, site last updated August 2016, https://www.epa.gov/climate-indicators/climate-change-indicators-ocean-acidity.

cause massive algal blooms, sometimes causing fish kills. Once algae die off, they are decomposed by microbes that deplete dissolved oxygen. In addition, warm water is less dense and holds less oxygen than cold water. A dissolved oxygen content less than 2 mg of O_2 per liter is called "hypoxic" (hypoxia is inadequate oxygen in tissue) and is generally bad for fish and shellfish. Oxygen depletion is spreading with a potential serious effect on fisheries.[21]

Glaciers, ice caps, and sea ice melt

Chapter 5, How Do We Know Global Warming Is Real?, described in detail the melting of glaciers, ice caps, and sea ice that are already occurring. The principal risk associated with ice melting is an increase in sea level worldwide. Other consequences are loss of habitat for animals living in the Arctic and Antarctic, such as polar bears and penguins, and the effect on indigenous populations that are being forced to relocate. In 2019 there was a news report that 50 hungry polar bears had invaded the town of Belushya Guba in Novaya Zemlya, an archipelago in the Barents Sea, northeast of mainland Russia. The sea ice where they normally foraged for seals has melted so the bears were seeking food in the town dump.[22] Belushya Guba is home to about 2000 persons, mostly Russian military. In 1961, it was the site of the airburst detonation of the "Tsar Bomba," the largest nuclear weapon ever detonated, with an explosive yield equivalent to 50 million tons of trinitrotoluene.

Polar ice ranges from 2−5 m thick (Arctic) to 2000−4000 m thick (Antarctic). The Greenland ice sheet ranges from a few meters thick at its fringes to thousands of meters at its thickest parts. To form these ice layers took thousands of years. There is concern that once substantial melting occurs, the process will become irreversible and cannot be stopped (Photo 7.2).

---

[21] Karen Limburg, "The Ocean is Losing its Breath—and Climate Change is Making It Worse," *SciTechConnect*, November 10, 2016, http://scitechconnect.elsevier.com/ocean-losing-breath-climate-change-worse/.

[22] The World at a Glance, "Novaya Zemlya, Russia: Polar Bears Invade," *The Week*, February 22, 2019, p. 9.

Photo 7.2 Hubbard Glacier, Alaska.

Subsidence occurs and permafrost melts

Subsidence can be caused by global warming, leading to structural damage or flooding. In Alaska, rising temperatures are thawing permafrost, causing the ground to subside, in some places up to 15 ft.[23]

Subsidence may be due to causes other than global warming, for example, by oil or groundwater extraction. Areas of Long Beach, California subsided more than 20 ft. in the 1940s due primarily to oil extraction. There were problems of flooding in the port areas. Later, water injection stopped the subsidence and in some areas raised the ground 2 ft. In southern Louisiana and along the Gulf Coast the land is subsiding and coasts are eroding. Parts of Houston are sinking at a rate of 2 in. or more per year, mostly due to pumping oil and water from under the city. Subsidence could make coastal areas all the more vulnerable to sea level rise and storm surge.

---

[23] Glick, *National Geographic*, p. 28.

Russia has some of the largest cities in the Arctic. In Norilsk, 60% of the buildings have been damaged by permafrost thaw, and 10% of the houses have been abandoned.[24]

A potentially much more serious issue is the prospect that a large permafrost melt could release methane from decaying organic matter trapped under the ice. Methane is a potent greenhouse gas, as mentioned previously. Ironically, this warning was sounded decades ago by a Russian scientist named Sergey A. Zimov living in northern Siberia. Zimov is a geophysicist who specializes in arctic and subarctic ecology. Originally his work was ignored, but now is receiving new interest (Textbox 7.3).

TEXTBOX 7.3

The town of Chersky, near the Kolyma River in northern Siberia is the home of 63-year-old Sergey A. Zimov. Zimov has observed that warming is taking place and the soil no longer freezes during the winter. His excavations indicate that the permafrost is much thicker here, compared to the typical depth of 1 m (3 ft.) found elsewhere. While sounding the alarm, Zimov also has a radical solution in mind. He thinks that by restoring the landscape to its prewarming condition, he can prevent further melting. This would require removing trees and bringing in livestock to forage and stir up the soil, enabling the permafrost to freeze. Recognizing that this is a long-term project, he has been joined by his son Nikita to carry on the work. It is interesting that here, 7000-plus miles from Hawaii, another father-son team is dedicating their lives to understanding global warming.

The worry is that potentially there is more greenhouse gas in permafrost than in all of the world's remaining fossil fuels. No one can say with certainty how much of it could be released as the Arctic warms. Temperatures in the Arctic continue to increase twice as fast as the rest of the world, according to the latest U.S. government climate report.[25]

---

[24] The Last Word, "When the Ground Melts," *The Week*, June 15, 2018, p. 37−38.

[25] Scott Pelley, "Siberia's Pleistocene Park: Bringing Back Pieces of the Ice Age to Combat Climate Change," *60 Minutes*, aired March 31, 2019, https://www.cbsnews.com/news/siberia-pleistocene-park-bringing-back-pieces-of-the-ice-age-to-combat-climate-change-60-minutes/.

Deserts and tropics expand

Desertification is a real risk of an overheated earth. Some areas could become uninhabitable. In parts of the world large populations eke out a living by subsistence agriculture. Their survival is entirely dependent on the weather. Extreme heat and drought dry up water supplies and destroy crops and livestock. Examples include areas such as in the Middle East and North Africa.

Species migration and extinction

Biologists know that every species has an optimum climate range for reproduction and survival. It depends not only on temperature, but also on rainfall and seasonal changes, including winter freezing. For example, as the earth warms the optimum ranges move to the north in the northern hemisphere. Historically, severe climate changes have occurred slowly over millennia, allowing time for plants and animals to adapt to changing conditions. Likewise, from the historical record we know that rapid change (such as the catastrophic impact of a large meteor) has led to massive species extinction.

In May 2019, the United Nations released a new report from the Intergovernmental Science Policy Platform on Biodiversity and Ecosystems Services (IPBES). It was compiled by 145 experts from 50 countries. The main conclusion is that over 1 million species of plants and animals are facing extinction due to human activity. The causes are described as farming and land-use practices that are destroying habitat, overfishing, global warming due to greenhouse gas emission, land and water pollution, and the spread of invasive species, including bacteria, insects, and plants.[26] In California, the outlook for the famed Joshua trees in Joshua Tree.

National Park is reported as "bleak." Hotter temperatures and less rainfall could spell the end of this species that dates to the Pleistocene era that ended 12,000 years ago.[27] Photos 7.3 and 7.4.

---

[26] Associated Press, "Nature in Deep Trouble, Report Says," *Los Angeles Times*, May 7, 2019, p. A4. See also United Nations, "UN Report: Nature's Dangerous Decline 'Unprecedented'; species extinction rates 'accelerating,'" May 6, 2019, https://www.un.org/sustainabledevelopment/blog/2019/05/nature-decline-unprecedented-report/.

[27] See "For a Park's Joshua Trees, Time May Be Running Out," *Los Angeles Times*, August 11, 2019, p. B3. Article based on a study by Lynn Sweet, University of California at Riverside Center for Conservation Biology.

Photo 7.3 Live Joshua Trees, Joshua Tree National Monument.

Photo 7.4 Dying Joshua Trees, Joshua Tree National Monument.

Some scientists have postulated that we are now living in the "Sixth Mass Extinction," a period some call the "Anthropocene Era," where as much as 75% of all species vanish from the earth. The cause is human destruction of habitat and human-caused greenhouse gas emissions.[28]

Many sources have reported the demise of a diminutive rodent, the Bramble Cay melomys, *Melomys rubicola*, considered the first (known) example of an extinction caused by global warming. An article noted that "Australian officials announced what is believed to be the first species of mammal to go extinct because of climate change brought on by human activity." The rat-like Bramble Cay melomys lived on a small sandy island at the northern end of Queensland's Great Barrier Reef before rising sea levels and repeated storm surges inundated its low-lying habitat. No trace of the rodent has been found since the last survey of its former habitat. This prompted Australia's environment ministry to finally place the *M. rubicola* on its "extinct" list.[29]

In the Grand Canyon, species are struggling to survive. As temperatures rise, many of the Grand Canyon's native species will need to move north. Bighorn sheep are particularly sensitive to climate. The spruce and fir forests that delight visitors and adorn the canyon's North Rim are vulnerable to drought and heat.[30]

Frequency and severity of storms

In 2013, a storm in southern England caused a five to 6-m rise in the sea level near coastal areas.[31] Storms like that have ravaged low-lying coastal areas. We certainly do not want any more storms such as hurricane *Katrina*, with over 1800 deaths and $81 billion in damages, or the more recent (2018) hurricane *Michael*, with a death toll of 47 persons and a cost of $25 billion.

---

[28] Briefing: "The Sixth Mass Extinction," *The Week*, February 22, 2019, p. 11. See also Elizabeth Kolbert, *The Sixth Extinction: An Unnatural History* (New York, NY: Henry Holt and Company, LLC, 2014).

[29] Earthweek: A Diary of the Planet, *Earthweek: Diary of a Changing World*, report from February 22, 2019, http://www.earthweek.com/. With thanks to Curt Abouch for bringing this to our attention.

[30] Steven Nash, Steven, "Grand Canyon's Next Century Will be Hotter," *Los Angeles Times*, January 23, 2019, p. A11.

[31] Sue Dawson, "How Europe's Coastal Cities Can Cope with Rising Sea Levels," *SciTechConnect*, March 16, 2017, www.scitecconnect.elsevier.com/coastal-cities-rising-sea-levels/.

When hurricane *Harvey* hit Texas in 2017, Houston's flood control system was supposed to handle a "100-year storm," defined as 13 in. of rain in 24 hours. The problem is that the "100-year storm" has already happened eight times in the last 27 years. *Harvey* was superpowered by warmer than usual Gulf waters. Besides massive rainfall, it caused a 15-foot-high surge in sea level along the coast. Within days, some parts of Houston had experienced 50 in. or more of rainfall—more than what is usual for a year. The result: tens of thousands of homes flooded and residents evacuated to shelters.[32]

Water temperatures in the Gulf of Mexico have steadily risen over the last 40 years. In the central gulf, the increase in average temperature has been 0.56°C (1°F), from 28.3°C to 28.9°C (83°F–84°F), while the average maximum temperature rise has been 1.7°C (3°F), from 31°C to 32.8°C (88°F–91°F).[33]

Dangerous weather conditions are certainly a risk from global warming. Meteorologists know that there is a link between warming and storm severity. Warm ocean water is what powers hurricanes, so global warming will increase their frequency and intensity. Sea level rise increases the likelihood of coastal flooding due to storm surge. Meteorologists believe that global warming was the reason *Harvey and Florence* were so destructive.

Rising seas and flooding caused by more severe storms compromises drinking water, the disposal of human waste, and increases the risk of water-borne diseases caused by pathogens such as bacteria, viruses, and protozoa.

Impact on agriculture, droughts, loss of cropland, and wildfires

Farmers need rain—but not too much or too little. Lack of rainfall can cause crops to be stunted or die, while too much can ruin crops before they can be harvested. Of these, the more serious one is lack of rain. With global warming, there is a prospect of certain regions of the world facing extended droughts that could displace millions of people. Increasing temperatures reduce the output of many food crops including

---

[32] Jenny Jarvis and Molly Hennessy-Fiske, "Texas Rainfall Tops Record," *Los Angeles Times*, August 30, 2017, p. A1.

[33] Climate Central, "U.S. Coastal waters temperature trends," September 8, 2016, https://www.climatecentral.org/gallery/graphics/coastal-water-temperature-trends.

grains such as rice and wheat. Temperature effects are in addition to other potential harm related to droughts or flooding.

The risk of wildfires in drought-stricken areas became abundantly clear in the long hot summer of 2018. Severe wildfires struck in four continents: Oceania, Europe, Asia, and North America, ranging from Siberia to Australia. Europe, Sweden, Greece, and the United Kingdom had major fires. There were fires within the Arctic Circle that were difficult to extinguish for lack of water. Particularly hard-hit were the U.S. western states and British Columbia. In California, more than 8000 fires burned 1.8 million acres, caused over $12 billion in insurance claims, and left 104 people dead. More than 18,000 structures were destroyed in the Camp Fire, the most destructive wildfire in California's history, when 85 people died. These 2018 fires broke all-time records for the state, both in terms of fire size and property destruction.[34]

In early 2020, wildfires had been raging in Australia since June 2019, with no end in sight. Australia's 2019/2020 fire season will be the most severe on record. Australia's fire season is during their summer months and is typically measured from September to March of the next year. It is estimated that up to 1.0 billion tons of greenhouse gases could be emitted by the end of the 2019–2020 season.[35] This is about twice the average emissions due to Australia's forest fires of 485 million tons per year for the past 10 years. It is almost double Australia's anthropogenic emissions (transportation, residential, commercial, and industrial emissions) of about 540 million tons per year. This is also larger than total emissions from all commercial aircraft in 2019 (915 million mt CO_2).[36]

The cause was prolonged droughts and unusually high temperatures. In early 2020, 18.6 million Ha (186,000 km^2) had burned or was burning. This is an area about twice the size of Hungary or Portugal. The damage included 6000 buildings, 30 human deaths, and an estimated 1 billion animals killed, many of them rare or endangered species. The Australian fires

---

[34] National Interagency Coordination Center (NICC), *2018 Statistics and Summary* (Boise, ID: NICC, 2018), https://www.predictiveservices.nifc.gov/intelligence/2018_statssumm/2018Stats&Summ. html.

[35] Denise Chow, "Australia Wildfires Unleash Millions of Tons of Carbon Dioxide," *NBC News*, January 22, 2020, https://www.nbcnews.com/science/environment/australia-wildfires-unleash-millions-tons-carbon-dioxide-n1120186.

[36] Akshat Rathi and Laura Lombrana, "Australia Fires Likely Emitted as Much Carbon as All Planes," *Bloomberg News*, January 20, 2020, https://www.bloomberg.com/news/articles/2020-01-21/australia-wildfires-cause-greenhouse-gas-emissions-to-double.

have so far burned more than 20 times the area of the 2018 California wildfires (800,000 Ha) or the 2019 Amazon wildfires (900,000 Ha).[37]

Elevated CO_2 means more photosynthesis and bigger leaf areas. This could mean more agricultural productivity, an increase in insect consumption, and plants would absorb less nitrogen from the soil. Farmers would use more nitrogen, creating a problem of runoff. When CO_2 increases, other elements decrease in crops, such as nitrogen, potassium, phosphorus, and others. Pests spreading to new areas would affect agriculture.

There are cascading effects:

- Plant suitability drops.
- Insects eat more.
- Yields decline.
- Altered nutritional status of parasites (that control insects).

Population growth by 2025 is anticipated to create large areas with over 100 persons per square kilometer. It is likely that by 2020 crops in China will not feed its population. Reductions in food supply will stimulate migrations and conflicts and bring about the risk that governments will nationalize supplies. This will hamper the ability to render aid to countries facing starvation.[38]

Health problems will be more severe

Global warming is a major threat to human health around the world. Less prosperous, less developed countries are more vulnerable. According to the *Lancet*, global warming compounds many of the problems faced by communities, such as providing housing, food supplies, water quality and supply, and preventing exposure to toxins, diseases, and other health hazards. Increased stress can affect mental health as well as physical health.[39]

Today, there are already billions of people who suffer from diseases or die because they do not have access to clean water for drinking and

---

[37] "Australia fires: A Visual Guide to the Bushfires and Extreme Heat" *BBC News*, December 31, 2019. Retrieved January 21, 2020, https://www.bbc.com/news/world-australia-50585968.

[38] John Trumbull, Professor, Department of Entomology, University of California at Riverside, personal communication in a lecture at the Beckman Center, Irvine, CA, November 7, 2018.

[39] David Introcaso, "Climate Change Is the Greatest Threat to Human Health in History," *The Lancet Countdown on health and climate change from 25 years of interaction to a global transformation for public health*, December 19, 2018.

Photo 7.5 Rural Ethiopian village.

hygiene or have poor sanitation. Many are infected by tropical diseases that are water-borne or hygiene-related. Infectious diseases transmitted by insects or through contaminated water spread more rapidly. Many insects are more active and breed more rapidly at higher temperatures. Warmer weather increases the spread of diseases carried by mosquitos, fleas, ticks, and rodents. Some infectious diseases will spread to previously unaffected geographic areas, for example as the tropics move northward. Warmer winters are resulting in an earlier annual onset of Lyme disease in the United States and a much higher rate of occurrence (Photo 7.5).

New medical reports state that "global warming makes us sicker." Without action to reduce greenhouse gas emissions, climate change will result in the early deaths of more than one-quarter million persons per year around the world, between the years 2030 and 2050. Deaths will result from lack of adequate nutrition, infectious and microbial disease, and heat exposure.[40]

Some studies will no doubt be controversial. A commission convened by *Lancet* Journal and consisting of 43 public health specialists from 14 countries is linking obesity, poor nutrition, and climate change. It argues that one problem cannot be solved without considering the other two. It links obesity to poor diet (processed food, too much red meat, etc.)

---

[40] Melissa Healy, "Doctors Weigh in on Global Warming: It Makes us Sicker," *Los Angeles Times*, January 24, 2019, p. A2.

In addition, farming methods to produce "manufactured" foods require more fossil fuels (for energy and fertilizer, etc.). Also, to reduce emissions, there is a need for better transportation and food storage, to reduce waste.[41]

Could global warming cause a financial crisis or some other financial problem?

There are a number of connections between the energy industries, banks, and major investors that could translate problems in energy industries into financial problems. This is not a discussion of what will happen, but what could happen.

So far, we have not had the equivalent of a 9/11 or Pearl Harbor event related to global warming that would cause a sudden and dramatic shift from fossil fuel use to renewables, and greater conservation and energy efficiency. It is possible for climate deniers and others who are trying to preserve the status quo to avoid taking actions to deal with global warming. This will not make the problem go away. It is possible that delays in taking action today will cause more disruptive changes in the future, especially if some precipitous event forces politicians, industry leaders, and the general public to suddenly realize that global warming is a real threat that needs to be dealt with more aggressively.

At present about one third of global financial securities are tied to fossil fuels from energy companies to equipment suppliers, and to banks and investment funds lending to or investing in companies in the fossil fuels industry. A "panic-driven decarbonization" could have severe economic repercussions, leading to rapid divestments and stranded assets. This is discussed in greater detail in Chapter 16, The Way Forward.[42]

---

[41] The Lancet, "Global Syndemic of Obesity, Undernutrition, and Climate Change: The Lancet Commission Report," January 27, 2019, https://www.thelancet.com/commissions/global-syndemic.

[42] Adam Tooze, "Why Central Banks Need to Step Up on Global Warming," *Foreign Policy Magazine*, July 20, 2019.

National security implications

Military bases and installations are negatively affected by extreme weather, flooding, and wildfires. In the United States, some of the most vulnerable facilities are along the coast, such as the U.S. Navy base in Norfolk, Virginia, described earlier. It is interesting to note that while the U.S. government denies that global warming is happening, the Department of Defense has quietly gone about surveying coastal facilities and making plans to raise them or relocate them to higher ground. Furthermore, the military is asked to respond to natural disasters such as *Hurricane Harvey* in 2017. An increase in the number and severity of climate change disasters such as storms and flooding will put additional burdens on the military.

U.S. Secretary of Defense Chuck Hagel said, "Climate change is a threat multiplier." For example, increasing problems associated with water scarcity, food shortages, and civil unrest could lead to conflict.[43] Some areas could become uninhabitable leading to mass migrations and threats to border security and stability in some countries. There could be conflicts over depleted resources such as food and water. Instability and increased violence create opportunities for terrorist activity.

There are also budget issues. Addressing natural disasters consumes funds needed for training, facilities, and equipment. The government's cost associated with addressing climate change and mitigating its effects could reduce military funding.

Global warming is a problem for those concerned about the threat from terrorism. Global warming is destroying cropland and grazing land in developing countries. Fewer young men will be able to support themselves as small farmers in the countryside. They will flock to already overcrowded cities looking for jobs that do not exist in the numbers required. This trend will be aggravated by population increases in these countries. The unemployed will be easy recruits for terrorist organizations who will pay them something for their services.

---

[43] Chuck, "Hagel, "Secretary of Defense Speech," *Conference of Defense Ministers of the Americas*, Arequipa, Peru, October 13, 2014, https://dod.defense.gov/News/Speeches/Speech-View/Article/605617/.

Migrations caused by climate change

Following the 2015 Paris Climate Agreement, British researchers studied how global warming would affect the geographical ranges of close to 100,000 plants, animals, and insects.[44]

To date, the average global temperature has increased by 1.0°C, compared to a pre-Industrial Revolution baseline. With an increase in global temperature to 2°C, the British study indicates that 18% of insects, 16% of plants, and 8% of vertebrates could lose more than half of their global ranges. If the increase is limited to 1.5°C, the number of affected species is much less: 6% of insects, 8% of plants, and 4% of vertebrates could lose more than half of their geographical range. At the other extreme, if efforts to curtail global warming fail, when the average global temperature increase reaches 3°C, the study indicates that over 40% of insects and plants and about one-quarter of vertebrates will see their ranges shrink by over 50%. More detailed field studies indicate that temperature alone is not necessarily the determining factor.[45,46,47] Other studies document that some species are either unaffected or are adapting.[48]

What about *Homo sapiens*? A recent study reported in *Science* indicates that we can expect a rise in asylum applications as temperatures increase in countries impacted by global warming.[49] The tragic stories of migrants crossing the Mediterranean Sea seeking asylum in the European Union

---

[44] R. Warren, J. Price, E. Graham, N. Forstenhaeusler, and J. VanDerWal. "The Projected Effect on Insects, Vertebrates, and Plants of Limiting Global Warming to 1.5°C Rather Than 2°C, *Science* 360, no. 6390 (2018): 791–795.

[45] Kevin C. Rowe, Karen M. C. Rowe, Morgan W. Tingley, Michelle S. Koo, James L. Patton, Chris J. Conroy, John D. Perrine, Steven R. Beissinger, and Craig Moritz, "Spatially Heterogeneous Impact of Climate Change on Small Mammals of Montane California, *Proceedings of the Royal Society B: Biological Sciences* 282 (January 2015), https://doi.org/10.1098/rspb.2014.1857.

[46] Christy McCain, Tim Szewczyk, and Kevin B. Knight, "Population Variability Complicates the Accurate Detection of Climate Change Responses," *Global Change Biology* 22 (2016): 2081–2093, doi:10.1111/gcb.13211.

[47] Susana M. Wadgymar, Jane E. Ogilvie, David W. Inouye, Arthur E. Weiss, and Jill T. Anderson, "Phenological Responses to Multiple Environmental Drivers Under Climate Change: Insights from a Long-Term Observational Study and a Manipulative Field Experiment," *New Phytologist* 218 (2018): 517–529, doi:10.1111/nph.15029.

[48] Constance I. Millar, Diane L. Delany, Kimberly A. Hersey, Mackenzie R. Jeffress, Andrew T. Smith, K. Jane Van Gunst, and Robert D. Westfall, "Distribution, Climatic Relationships, and Status of American Pikas (*Ochotona princeps*) in the Great Basin, USA," *Arctic, Antarctic, and Alpine Research* 50, no. 1 (2018), doi:10.1080/15230430.2018.1436296.

[49] Anouch Missirian and Wolfram Schlenker, "Asylum Applications Respond to Temperature Fluctuations," *Science* 358, no. 6370 (2017): 1610–1614.

have been in the news. While many migrants seek to escape conflict, others flee because their farms are failing, they have no work, water supplies are inadequate, or they are starving.

Applications from 103 source countries for persons seeking asylum in the European Union were examined. The results indicated that when the temperature in the source country deviated from the moderate optimum ($\approx 20°C$), asylum applications increased rapidly. Applications could potentially increase from a yearly average of 351,000 per year to 449,000 or as much as over 1 million applications per year.

European and other developed countries traditionally have been able to absorb refugees fleeing conflict or drought, but their capacity to do so is not unlimited. Also, there is a time factor. Establishing refugee camps, setting up medical facilities, and developing job training and social programs take time. A danger is that even the best of intentions could be completely swamped by a massive influx of needy migrants, causing the aid system to break down and trigger humanitarian and national security crises in many parts of the world.

Tipping points: unanticipated changes can occur

According to a recent U.S. government report, "There is broad consensus that the further and the faster the earth system is pushed toward warming, the greater the risk of unanticipated changes and impacts, some of which are potentially large and irreversible." In addition, "Future changes outside the range projected by climate models cannot be ruled out and could result in unanticipated surprises."[50]

With increasing temperatures, there is an increased risk that some threshold or *tipping point* (a point beyond which some unanticipated and more severe action could occur) will be exceeded, leading to unpredictable climate effects.[51] Here are some examples:

---

[50] Wuebbles et al., *NCA4*.

[51] Jonathan Watts, "Domino-effect of climate events could move earth into a 'hothouse' state," *The Guardian*, August 7, 2018, https://www.theguardian.com/environment/2018/aug/06/domino-effect-of-climate-events-could-push-earth-into-a-hothouse-state. See also Will Steffen, et al., "Trajectories of the Earth System in the Anthropocene," *PNAS* 115, no. 33 (August 14, 2018): 8252–8259.

- Sea level increases could accelerate. It is possible that sea levels could rise by as much as 2.4 m (8.0 ft.) by the end of the century. This is not a forecast but a possibility of what could happen if there was sufficient melting of ice caps in Greenland and Antarctica. The possible collapse of the Antarctic ice sheet would increase the rate at which glaciers are melting. As the sea ice breaks away, glaciers are free to slide into the ocean.
- Melting of polar ice could decrease the amount of reflected sunlight, causing still more melting, sea level rise, and faster loss of ice.
- Ocean currents such as the Gulf Stream could slow down, causing major climate changes such as much colder winters in the United Kingdom and Scandinavia.
- The Arctic is warming at twice the global average rate. As mentioned above, melting permafrost could release large quantities of methane, a potent greenhouse gas, causing the earth's increasing temperature to accelerate, causing still more melting, releasing more methane, and so on. This is a classic example of "positive feedback," where a small change causes a much larger effect. Greater warming in turn causes further adverse effects that could spiral out of control. The East Siberia Arctic shelf contains over 1000 Gt of carbon. The release of a small percentage of this (say 50 Gt) would equal total greenhouse gas emissions in 2018.[52]

At present, it is not possible to forecast any of these events with certainty. These and other nonlinear effects remain as remote, but potentially devastating, possibilities.

In summary, the risks enumerated above are serious enough that we need to think carefully about possible solutions. What would it take to stop global warming?

---

[52] Nafeez Ahmed, "Seven Facts You Need to Know About the Arctic Methane Time Bomb, *The Guardian*, August 5, 2013.

CHAPTER 8

International efforts to address global warming

There is a long history of international efforts to understand the dynamics of the earth's environment. Some early investigations identified a potential problem associated with global warming from man-made emissions of CO_2.

Early efforts[1]

The year 1957, the same year that *Sputnik* was launched, was designated the International Geophysical Year and led to more funding and a more coordinated international approach to climate research. The International Global Atmospheric Research Program was established in 1967 to gather data for weather forecasting and also for climate research. That same year, it was calculated that doubling CO_2 concentration in the atmosphere would increase atmospheric temperatures about two degrees.

The U.S. National Oceanic and Atmospheric Administration (NOAA) was established in 1970 and became the world's leading source of funding for climate research. By the late 1970s, scientific opinion tended to conclude that global warming was a major climate risk. The U.S. National Academy of Sciences reported that it is credible that doubling CO_2 concentration in the atmosphere would increase the earth's temperature from 1.5°C to 4.5°C. The World Climate Research Program was also launched to coordinate international climate research.

The year 1981 was reported as the warmest year on record at that time. It was also reported that there was significant global warming since the mid-1970s. Later, Antarctic ice cores showed that CO_2 concentration and the earth's temperature tracked each other up and down together during past ice ages.

---

[1] American Institute of Physics, The Discovery of Global Warming, Timeline (Milestones), February 2019, https://history.aip.org/climate/timeline.htm.

Reaching Net Zero.
DOI: https://doi.org/10.1016/B978-0-12-823366-5.00008-7

© 2020 Elsevier Inc.
All rights reserved.

In 1987, the Montreal Protocol of the Vienna Convention imposed international restrictions on the emission of ozone-destroying gases. This was to address the hole in the ozone layer. This is perhaps the first example of international cooperation to solve a global environmental problem. The following year, in 1988, the Toronto Conference on the Changing Atmosphere called for limits on greenhouse gas (GHG) emissions. The Intergovernmental Panel on Climate Change (IPCC) was established.

The Intergovernmental Panel on Climate Change

The first IPCC report was issued in 1990. It concluded that global warming was occurring and future warming was likely. In 1992, a conference in Rio de Janeiro produced the UN Framework Convention on Climate Change. In 1995, the Second IPCC report linked man-made GHG emissions to global warming and predicted significant global warming in the future. Reports of breaks in the Antarctic ice shelves and other signs of global warming started to influence public opinion.

In 1997, an international conference produced the Kyoto Protocol that set targets for industrial nations to reduce their GHG emissions. More than 150 countries ratified the protocol but not the United States. The United States dropped out in 2001, with President G.W. Bush claiming it would hurt the United States economy.

The Third IPCC report was issued in 2001 stating that global warming is unprecedented since the end of the last ice age. At a meeting in Bonn, many countries (but not the United States) started to develop the means of working toward the Kyoto targets of 1997. In 2005, the Kyoto treaty went into effect and was signed by the major industrial nations except for the United States. This treaty led to efforts by several countries to work toward limiting their GHG emissions.

About this time, China overtook the United States as the world's largest emitter of CO_2, the main GHG.

The Fourth IPCC report was issued in 2007 and warned of the serious effects of global warming. A year later, climate scientists recognized that even if all GHG emissions could be stopped immediately, global warming would continue. This is due to the fact that CO_2 released to the atmosphere remains there for hundreds of years.

International efforts to address global warming   101

In 2009, a conference in Copenhagen failed in its attempt to negotiate binding agreements controlling GHG emissions.

The Fifth IPCC report was issued in 2014. This report confirmed that GHG emissions from man-made sources are the main cause of global warming. This report also highlighted the risks associated with global warming and associated climate change. In addition, the report stated that the means are available now to limit global warming. This report established a target of trying to stabilize the earth's temperature increase to no more than 2.0°C above preindustrial levels.

In October, 2018, the IPCC was invited to issue a special report on the impact of global warming of 1.5°C above the average preindustrial global temperature level. This report concluded that GHG emissions would have to be reduced to net zero by 2050 to avoid a temperature rise of 1.5°C. This report is discussed in more detail in Chapter 9, What Would It Take to Reach Net Zero?.

The Paris Agreement[2]

In 2015 the Paris Agreement was signed whereby all nations including the United States agreed to set targets for their GHG emissions and report their progress periodically. This was a breakthrough international agreement dealing with climate change. The main goal of the Paris agreement was to limit the global average temperature rise above preindustrial levels to be less than 2.0°C, and preferably under 1.5°C.

That same year, 2015, the global mean temperature had reached 14.9°C (59°F), the warmest in thousands of years. The level of CO_2 in the atmosphere exceeded 400 ppm for the first time in millions of years.

In 2017, newly elected President Trump announced that the United States would withdraw from the Paris Agreement since the limits on GHG emissions would undermine U.S. economic growth. The Trump administration has formally notified the United Nations that the United States is withdrawing from the Paris Agreement. This withdrawal will be completed after a 1-year waiting period has elapsed. The United States

---

[2] Fiona Harvey, "How the UN Climate Panel Got to the 1.5°C Threshold Timeline," *The Guardian*, October 7, 2018, https://www.theguardian.com/environment/2018/oct/08/how-the-un-climate-panel-ipcc-got-to-15c-threshold-timeline.

did agree to continue participating in international climate change negotiations for now.

In the United States, governors of several U.S. states formed the United States Climate Alliance to continue advancing the objectives of the Paris Agreement. In 2018, Governor Jerry Brown of California signed Senate bill SB100 committing California to be 100% emissions-free for electricity generation by 2045. This has been followed by Hawaii and other states. California has moved ahead rapidly with new wind and solar projects. In June 2019, the state set two records: the most solar power ever flowing on the main electric grid and the most solar power taken off line because it was not needed. This is a positive sign for rate payers. The state needs to eliminate expensive fossil fuel power purchase agreements so cheaper renewable energy can be used.[3] In another positive sign, the Mayor of Los Angeles vetoed the city's Department of Water and Power's plan to spend billions of dollars to rebuild three natural gas—fired generating stations located along the coast. Los Angeles has also moved away from coal, divesting from the Navajo coal-fired power plant in Arizona and announcing plans to stop purchasing power from Utah's Intermountain plant by 2025.[4]

Another significant global organization is the *Global Covenant of Mayors for Climate and Energy*. This organization counts over 9000 cities representing over 800 million people worldwide with a mission to accelerate ambitious, measurable climate and energy initiatives to meet or exceed the Paris Agreement objectives.[5]

Intergovernmental Panel on Climate Change special reports

In 2019, the IPCC issued two special reports. The first dealt with *Climate Change and the Land* and the second dealt with the *Ocean and Cryosphere in a Changing Climate*. In addition to the IPCC reports, the

---

[3] Sammy Roth, "Too Much Solar Power May Brighten Up Ratepayers' Day," *Los Angeles Times*, June 6, 2019, p. C1.

[4] Sammy Roth, "L.A. to Scrap Plans for Gas Plants: Mayor Changes Course, Saying City Must Move Toward Renewable Energy and Improve Air Quality," *Los Angeles Times*, February 12, 2019, p. C1.

[5] The Global Covenant of Mayors for Climate & Energy, "This is a Powerful and Historic Response to Climate Change," accessed February 8, 2020, https://www.globalcovenantofmayors.org/about/.

United Nations Environmental Program (UNEP) issues an annual *Emissions Gap Report* summarizing the results of current environmental studies and estimates of future GHG emissions. This estimate is compared to the goals stated in the Paris Agreement, the "Nationally Determined Contributions" (NDCs) stipulated by each party to the agreement. The purpose is to determine the gap between current emissions pathways and what is required to meet the established goals. The *Emissions Gap Report* 2018 concluded that unless the NDCs established by the Paris Agreement are increased before 2030, exceeding the 1.5°C goal is unavoidable. The tenth *Emissions Gap Report* issued in 2019 concluded that countries have collectively failed to stop the growth in global GHG emissions with no sign of emissions peaking in the next few years.[6] The emissions gap is large. In 2030, annual emissions need to be 15 GtCO$_2$eq lower than current unconditional NDCs to meet the 2°C goal and 32 GtCO$_2$eq lower for the 1.5°C goal. However, with only current policies, GHG emissions are estimated to be 60 GtCO$_2$eq in 2030. Even if current unconditional NDCs are fully implemented, there is a 66% chance that warming will be limited to 3.2°C by the end of the century.

The next major IPCC report will be the Sixth Assessment Report. In September 2020, experts from 65 countries, the Working Group, will meet in India to start preparing the first draft of this report. The final report is due in July 2021.

History of Intergovernmental Panel on Climate Change global warming objectives[7]

What is the origin of the IPCC's objective stated in the Paris Climate Agreement of limiting global warming to less than 2°C with a further objective to limit warming to 1.5°C? The 2°C limit was initially proposed by economist Richard Nordhaus in 1975 based upon his intuition that "more than 2°C above preindustrial levels would take climate

---

[6] UN Environment Programme, *Emissions Gap Report 2019*, November 26, 2019, https://www.unenvironment.org/resources/emissions-gap-report-2019.
[7] CarbonBrief, Rosamund Pearce, "Two degrees: The History of Climate Change's Speed limit," December 8, 2014. A graphic that shows the history of international discussions from William Nordhaus (1975) to the Cancun agreements (2010), to "hold the increase in global average temperature below 2°C." https://www.carbonbrief.org/two-degrees-the-history-of-climate-changes-speed-limit.

outside the range of observations that have been made over the last several hundred thousand years."

In 1996 the European Council of environmental ministers was the first major political organization to formally endorse this limit stating that global average temperatures should not exceed 2°C above preindustrial levels. In 1997 193 countries signed the world's first agreement on GHG emissions, the Kyoto Protocol. No temperature limit was specified at the time. The United States did not ratify the Kyoto Protocol.[8]

Starting in about 2008, the Alliance of Small Island States (AOSIS) called for more ambitious limits and supported a limit well below 1.5°C to reduce the disastrous impact of rising sea levels.

In 2009 the Copenhagen Accord accepted the 2.0°C temperature limit as the central goal but said it would consider limiting the temperature rise to 1.5°C.[9] In 2010 the Cancun Agreements committed governments to limit global temperature increases below 2°C. This agreement also committed to review if the 2.0°C limit needed to be strengthened to limit the increase to 1.5°C.

Finally, in 2015, the landmark Paris Agreement was signed under which 195 countries supported the long-term goal to limit global warming to well below 2°C and pursue a goal of 1.5°C. At this conference, the IPCC was asked to submit a special report in 2018 on the effects of limiting global temperatures to 1.5°C above preindustrial levels and discuss pathways leading to this objective. This report, titled *The Intergovernmental Panel on Climate Change (IPCC) (October 8, 2018) Special Report on Global Warming of 1.5°C (SR1.5)*, is frequently abbreviated as SR15.

---

[8] Carla Tardi, "The Kyoto Protocol," *Investopedia*, September 26, 2019, https://www.investopedia.com/terms/k/kyoto.asp.

[9] The Copenhagen Climate Change Conference, December 2009. This was a meeting involving over 100 world leaders. It produced the Copenhagen Accord, one of the first major documents to express a political intent to constrain carbon emissions. https://unfccc.int/process-and-meetings/conferences/past-conferences/copenhagen-climate-change-conference-december-2009/copenhagen-climate-change-conference-december-2009.

PART II

CHAPTER 9

What would it take to reach net zero?

The previous chapters presented the hard physical evidence that proves the earth's temperature is rising due to increased greenhouse gas emissions caused by humans. In 2018, total global greenhouse gas emissions reached a new high of 55 $GmtCO_{2eq}$/year. Of this total, approximately two-thirds or 37 $GmtCO_2$/year was carbon dioxide, largely from combustion of fossil fuels (see Fig. 9.1).

Total emissions include the impact of land-use changes, mainly the impact of deforestation to create more land for agriculture. Land-use changes are not a source of greenhouse gases but reduce sequestration of CO_2 when trees are eliminated, causing more CO_2 to be absorbed by the atmosphere or ocean. Emissions due to land-use changes have been flat since at least 1970. However, this could change. Forests continue to be destroyed. The world's growing population will need more land for food production and lead to increasing deforestation unless actions are taken to prevent this.

Intergovernmental Panel on Climate Change alternative scenarios

Fig. 9.2 shows a summary of the various Intergovernmental Panel on Climate Change (IPCC) alternative scenarios from its 2018 report along with what needs to be done to limit global warming to 2.0°C or less. The shaded areas represent the range of uncertainty in the estimates. While the IPCC and other international organizations would prefer that global warming be limited to 1.5°C, they recognize that a 2.0°C increase is a more realistic target.

There are three alternatives. The first is what might happen if we do nothing or not enough to stop global warming. The earth's temperature rise could exceed 4.0°C by 2100. This alternative would also increase the

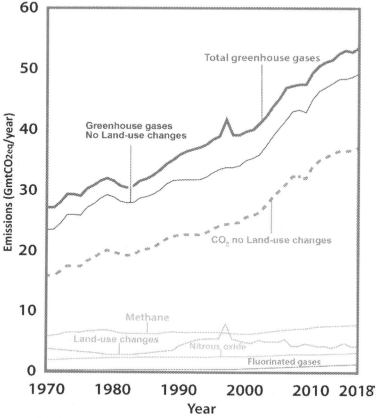

Figure 9.1 Global emissions of CO_{2eq} in 2018.[1]

risk that some nonlinear effects or some tipping point is reached with unforeseen consequences.

The second alternative assumes that most or all of the countries that signed the Paris Agreement fulfill their pledges and meet the CO_2 reductions promised. This would reduce but not stop global warming. The 2.0°C limit would still be exceeded. Pledges to reduce greenhouse gas emissions made by countries signing the Paris Agreement, if achieved, are currently not sufficient to limit global warming to 1.5°C. In addition,

---

[1] See UN Environmental Programme, 2018 Gap Report, Figure 2.3 in Final report pdf, https://www.unenvironment.org/resources/emissions-gap-report-2018. Used with permission. UN Environment Programme, Pinya Sarasas, Programme Officer.

What would it take to reach net zero?

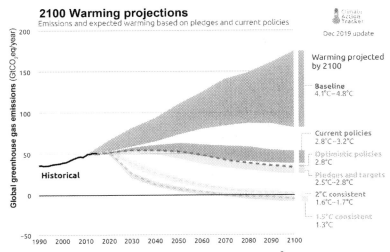

Figure 9.2 IPCC global warming scenarios and projections.[2]

there is no enforcement mechanism. If pledges are implemented, it is estimated that greenhouse gas emissions would be 52–58 GmtCO$_{2eq}$/year in 2030, far short of required reductions. In addition, to make matters worse, on June 1, 2017, the Trump administration stated its intention to withdraw the United States, the world's second largest source of greenhouse gases, from the Paris Agreement.

The third scenario assumes that there is a rapid, effective global response and the countries accounting for the vast majority of CO$_2$ emissions do what it takes to get to net zero emissions of CO$_{2eq}$ to stop global warming.

If the national pledges made under the Paris Agreement are not sufficient, what would it take to stop global warming?

This is the question asked by the United Nations Framework Convention on Climate Change (UNFCCC) following the adoption of the Paris Agreement in 2015. The IPCC was invited to provide a special report on the impact of global warming of 1.5°C above preindustrial levels. The report also discusses various greenhouse gas emissions pathways to achieve 1.5°C. This report is referred to as *The Climate Science Special*

---

[2] Climate Action Tracker, "2100 Climate Predictions," 2100 warming projections from IPCC, based on current pledges and policies, site last updated December 2019, https://climateactiontracker.org/global/temperatures/.

Report, or SR1.5 (October 2018). The report was to not only assess what a 1.5°C warmer world would look like but to also discuss the different pathways by which global temperature rise could be limited to 1.5°C. According to SR1.5, to limit global warming to less than 1.5°C, CO_2 emissions would have to be reduced to net zero by 2050.[3]

Man-made global warming has already reached about 1.0°C above preindustrial levels. The earlier limit of 1.5°C suggested by the IPCC is less than half a degree above where we are now.

The U.S. government also issued a special report on global warming in November 2018. According to the U.S. government's *The Climate Science Special Report,* without major reductions in emissions, the increase in annual average global temperature relative to preindustrial times could reach 5°C (9°F) or more by the end of this century.[4] With significant reductions in emissions, the increase in annual average global temperature could be limited to 2°C (3.6°F) or less.

There are a number of uncertainties in making a forecast for global warming. Forecasts are usually presented as a range of possible outcomes or as a most likely estimate with a band or showing the range of possible high and low estimates due to uncertainties in the data or the analysis and forecast.

Countries signing the Paris Agreement pledged to reduce greenhouse gas emissions as stated in individual "Nationally Determined Contributions." Even if all these pledges are fulfilled, it will not be enough to get greenhouse gas emissions to net zero. In our opinion the earth's temperature would likely exceed 3.0°C by the end of this century even if these pledges are fulfilled.

What would it take?

According to the IPCC, complete decarbonization is required, reducing man-made greenhouse gas emissions to net zero. As noted in Fig. 3.4, in the preindustrial age, the amount of carbon entering the atmosphere equaled the amount absorbed by the earth. This situation is referred to as "net zero." The maximum temperature reached is

---

[3] V. Masson-Delmotte, et al., IPCCSR1.5.
[4] Wuebbles et al., *NCA4.*

> **TEXTBOX 9.1**
> Net zero is when greenhouse gas emissions (CO_2, CH_4, and others) are reduced such that no more are added to the atmosphere. Any CO_2 emissions that cannot be reduced have to be offset by sinks such as forests or plant life that absorb excess CO_2, or by artificial means to remove carbon from the atmosphere. Achieving net zero will stabilize but not reduce the earth's temperatures until greenhouse gases already in the atmosphere gradually dissipate naturally. This could take hundreds of years.

determined by cumulative excess global anthropogenic (human-caused) CO_2 emissions up to the time that net zero is achieved. The ongoing effects of global warming will continue for a long time even after net zero is achieved (Textbox 9.1).

We not only have to eliminate CO_{2eq} emissions and achieve net zero, but we also need to stop and hopefully reverse the destruction of natural sinks such as large forests. The destruction of carbon sinks, mainly the clearing of forests to create farm and pasture land, accounts for about 13% of greenhouse gas emissions.

Since about 1880, an estimated 2200 $GmtCO_{2eq}$ has been dumped into the atmosphere, mainly due to fossil fuel use (Appendix 2). Almost all of the CO_2 is still in the atmosphere. If global warming is to be limited to 1.5°C, then additional CO_2 emissions have to be limited to about 420 $GmtCO_{2eq}$, our carbon budget. The carbon budget is an approximation of the amount of CO_{2eq} that can be added to the atmosphere before a specific temperature is exceeded, in this case 1.5°C. Refer to Chapter 14, Action Plan: Efficiency, Power, Transportation, and Land Use, for discussion of carbon budgets.

Are the Intergovernmental Panel on Climate Change scenarios realistic?

Table 9.1 lists the scenarios presented in IPCC SR1.5. This report also notes that there is no single definitive pathway to limiting global warming to 1.5°C. Several alternatives are explored.

Table 9.1 Summary of IPCC scenarios.

Scenarios

P1 A decline in energy demand, improved efficiency, rapid decarbonization. Afforestation for some carbon removal. Get to net zero by 2050.

P2 Similar to P1 with some industrial carbon removal to accommodate limited residual fossil fuel use.

P3 Less aggressive decarbonization and some reduction in energy demand. Heavier industrial carbon removal to offset residual fossil fuel use and remove any overshoot of CO_2 in excess of what is allowed by the 1.5°C limit.

P4 Limited changes in energy use offset by improving technology and widespread use of industrial carbon removal.

SR1.5 evaluates two main alternatives. The first, called P1, involves taking actions to prevent global warming from exceeding the 1.5°C limit. To do this, we would have to get to net zero by 2050. The others (P2, P3, and P4) would allow global temperatures to rise above this limit for a few decades. These are referred to as "overshoot pathways." These pathways assume that it becomes practical technically and economically to remove large amounts of CO_2 from the atmosphere to eventually reduce global warming to 1.5°C. The problem is that CO_2 removal from the atmosphere at the huge scale required to reduce the earth's temperature has never been done.

If a large overshoot in temperature occurred, it could result in irreversible damage such as a collapse of a polar ice sheet followed by sea level rise even if the earth's temperature is eventually lowered to 1.5°C (Refer to Chapter 5).

Few if any of the efforts currently underway would be sufficient to limit global warming to less than 1.5°C. A very rapid escalation to expand current efforts and accelerate their implementation is required.

Scenarios P1, P2, and P3 are referred to as low or no overshoot scenarios while P4 is referred to as a high overshoot scenario, meaning that CO_2 emissions greatly exceed reduction targets and have some residual fossil fuel emissions that will require carbon removal to prevent temperature increases from exceeding the 1.5°C objective. Scenarios P2, P3, and P4 assume that it will become feasible to remove large amounts of CO_2 from the atmosphere in time to make a difference (see Fig. 9.3). This capability is used, after overshooting the carbon budget limit, to remove

What would it take to reach net zero?

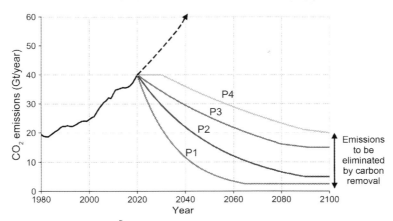

Figure 9.3 IPCC scenarios.[5]

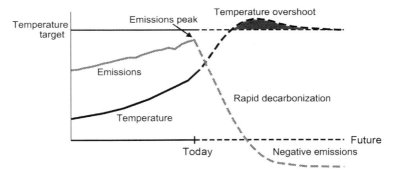

Figure 9.4 Example of overshoot and return to temperature target.

CO_2 from the atmosphere to get back to a CO_2 concentration consistent with the 1.5°C objective (see Fig. 9.4).

We question the feasibility of all four scenarios for the following reasons:

Except for P4, it is assumed that decarbonization will begin in 2020 with rapid reductions in CO_2 emissions. Proposed reductions are much more aggressive than what countries agreed to in the Paris Agreement—and even these reductions are not enforceable and are not being met by several countries including the United States.

It is very *unlikely* that an international agreement can be reached near-term under which all major CO_2 emitters agree to aggressive reduction

---

[5] V. Masson-Delmotte, et al., IPCCSR1.5.

targets and other actions required. Developing the required level of cooperation needed between the United States and China, the two largest CO_2 emitters, and other major developed and developing countries would take a long time if it is even achievable at all.

The last three scenarios assume that CO_2 removal of from 100 to 1000 $GmtCO_2$ is possible. CO_2 emissions will exceed the limit needed to keep warming below 1.5°C. This excess CO_2 would then have to be removed by unproven industrial-scale carbon removal facilities that are being developed. Even if it is proven to be feasible, it would take time to develop and deploy industrial-scale installations able to remove enough carbon from the atmosphere and store it underground. Even if this is possible, it is not unreasonable to assume that installations big enough to make a difference would require costly investments to build and large amounts of renewable energy to operate. Who would make this investment and who would pay to operate carbon removal equipment? Also, failed attempts to store radioactive wastes underground should provide a cautionary note. Would we be able to safely store all the CO_2 removed from the atmosphere forever?

In a recent paper, Professor James Hansen estimates that a 1.0 ppm reduction in the CO_2 in the atmosphere requires removal of 7.8 billion tons of CO_2. The cost of carbon capture used in his estimate was $113–$232 per ton of CO_2 removed from the atmosphere.[6] This is only the cost of CO_2 capture and does not include the additional cost of transporting the CO_2 and storing it underground. Using these estimates, Professor Hansen calculates that it would cost between $878 billion and $1.8 trillion to lower atmospheric CO_2 by 1.0 ppm. It is obviously a lot cheaper to avoid putting the CO_2 into the atmosphere in the first place.

CO_2 removed from the atmosphere might not have any economic value, especially if it is stored underground to sequester the CO_2. The cost of removing CO_2 would have to be paid by governments. If there was a high enough fee on carbon emissions, then revenue from carbon fees could be used to remove and store CO_2. This fee would obviously have to be higher than the cost of removing CO_2, at least $100/ton or higher.

---

[6] James Hansen, "Climate Change in a Nutshell: The Gathering Storm," December 18, 2018, http://www.columbia.edu/~jeh1/mailings/2018/20181206_Nutshell.pdf.

Carbon removal

Several methods are under development. One method, called carbon capture and storage, involves capturing CO_2 from a fossil fuel power station, removing the CO_2 chemically, and then permanently storing the captured CO_2 underground in depleted oil wells. When clean coal is discussed, it assumes that the CO_2 released is captured and stored. If successful, this method would reduce the amount of CO_2 released into the atmosphere from fossil fuel use. It would not remove any CO_2 already in the atmosphere. Removing CO_2 from power plant exhaust could be economical if there is a high enough carbon fee that could be avoided by removing and storing CO_2.

The second method, called direct-air capture and storage, involves the large-scale removal of CO_2 directly from the atmosphere followed by storing it underground. Some of the CO_2 could be injected into oil wells to increase the amount of oil recovered. This is referred to as enhanced oil recovery. A pilot plant is under consideration for the Permian Basin in Texas. It is assumed that the CO_2 injected into an oil formation will be captured underground forever for this to be an effective means of reducing CO_2 in the atmosphere. The CO_2 would have a value associated with the additional oil recovered in order to justify the cost of building and operating the CO_2 capture facility.

A third alternative is for the CO_2 removed from the atmosphere to be converted to a synthetic liquid fuel that can substitute for gasoline, diesel fuel, or aviation fuel. The production of synthetic fuels from captured CO_2 is discussed in Chapter 10, Energy Alternatives. Like biofuels, synthetic fuels do not permanently remove CO_2 from the atmosphere, since they recycle the CO_2 into fuel that will release CO_2 back into the atmosphere when the fuel is consumed. The benefit is that it avoids adding *additional* CO_2 (as would occur if regular fossil fuel was used), so long as the entire process is powered by electricity from renewable sources.

What is a more likely scenario?

If it is unlikely that aggressive global actions will be taken soon, then something like P4, but without carbon removal, is the most likely scenario. If the world is not willing to spend the effort and expense to aggressively reduce emissions, why should we think there will be a global effort to develop an unproven technology and spend the trillions of dollars to build and operate facilities to remove CO_2?

We should keep in mind that no matter what temperature is eventually reached, we have to do as much as we can to stop global warming now. As stated earlier, CO_2 put into the atmosphere stays in the atmosphere essentially forever on a human time scale. Ongoing CO_2 emissions will lead to further temperature increases with more severe permanent effects on our climate.

Are we too late already?

It is extremely unlikely that immediate, aggressive actions will be taken by the countries that are the largest emitters of CO_2. For the United States, it may have to wait for a change of administration before it will decide to remain in the Paris Agreement and support an international effort to address global warming. We think it likely that it will take 5 years or more for an adequate international effort to be agreed to and to start implementation, probably in 2025 or later.

Fig. 9.5 shows two possible pathways from the present to 2050. The first, highly improbable, is what would be necessary to achieve the IPCC

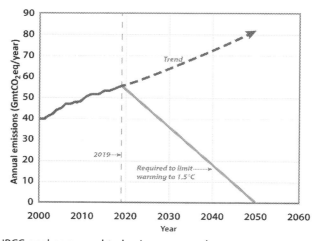

Figure 9.5 IPCC goal compared to business as usual.

goal of net zero by 2050. The second, is "business as usual," a continuation of the historic trend.

Why bother if we cannot limit the earth's temperature rise to $1.5°C$? We do not have a choice. Exceeding this limit means that temperatures would be higher than the objective, the effects of global warming would be greater, and the effort and cost to mitigate the effects of global warming will be higher. The longer we wait, the bigger the price we will have to pay.

Doing nothing is not an option

We will have to respond to global warming eventually by design or duress. If we continue without doing enough to stop global warming, we will reach some limit or experience some massive effect that will force us to recognize the dangers associated with unchecked global warming. To quote the late economist Herb Stein, "If something cannot go on forever, it will stop." We just do not know how or when. Even then, will we do what is needed?

We must keep in mind that global warming will continue to increase as long as CO_{2eq} emissions are positive, above net zero. At some point, temperature increases and associated climate impacts will become obvious and unacceptable.

As we have stated earlier, it is very likely that it is too late to limit the earth's temperature rise to $2.0°C$, the upper limit set in the Paris Agreement signed in 2015. By the time aggressive, global actions can be agreed to and implemented, the carbon budget associated with a $2.0°C$ temperature increase would be exceeded. It is also possible that, even then, we will not be able to organize and implement the efforts needed to decrease CO_{2eq} emissions to net zero to avoid even higher temperatures.

What then? The impact of continuing global warming will be uneven and inequitable. Some people and some regions will be adversely affected much more than others. At some point, will every country that needs to do something make the required commitment to reduce CO_2 emissions as needed? Or, will those least affected be reluctant to alter their lifestyles for the good of the planet? This is the "Tragedy of the Commons" under which some people and countries can and will take greater advantage of our "commons," the earth's biosphere (atmosphere, oceans, land, and living things) at the expense of the rest. It is probably part of human nature that we worry more about our own comfort and convenience today than the welfare of others, especially if it is in the future.

To get to net zero or as close as possible, aggressive actions would have to be taken now. Some positive things are happening. As stated in Chapter 15, Can It Be Done?, "the trend is our friend" but is not sufficient by itself to get to net zero in time. Without additional actions these positive trends are not enough to stop global warming:

- Global energy efficiency is improving as measured by the amount of energy needed to produce a given increase in GDP. Energy efficiency is improving by about 2.0%/year. More economic growth is taking less energy.
- In addition, the use of solar and wind to produce electricity is increasing. In 2019 Bloomberg New Energy Finance (BNEF) reported that renewables are now the cheapest form of new electricity generation across two-thirds of the world—cheaper than both new coal and new natural gas power plants. The cost of solar- and wind-produced electricity continues to decline, making renewable energy increasingly competitive with fossil fuels and nuclear power.[7]
- The production and use of electric vehicles are increasing, especially in China, the country responsible for the largest greenhouse gas emissions.
- The use of electricity from renewables is fundamentally more efficient than using fossil fuels. Fossil fuels have to be burned in an internal combustion engine or power plant to produce usable power or electricity that can be used for power or heat. This process is typically 30%—40% efficient. About 60% or more is waste heat carried off in cooling water, radiators, or exhausts.

What will happen if we do nothing?

We do know that increasing global temperatures will amplify the effects of global warming discussed in Chapter 6, How Do We Know

[7] Seb Henbest, et al., "New Energy Outlook 2019," *Bloomberg New Energy Finance*, https://about.bnef.com/new-energy-outlook/.

Man-Made CO_2 Is the Issue? The global demand for energy is increasing due to population growth and improving living standards in most countries. If we continue to obtain 80% or more of this energy from fossil fuels, CO_2 emissions will keep increasing. Atmospheric concentrations of CO_2 and other greenhouse gases will increase, leading to further increases in the earth's temperature. Remember that once CO_2 is added to the atmosphere, it takes hundreds of years for that level of CO_2 to be naturally absorbed by the earth.

In summary:

The earth's temperature will continue to increase. There will be record-breaking temperatures and more severe heat waves. As shown in Fig. 9.6, temperature increases due to global warming could reach from 2.5°C to 4.7°C by the end of this century, based upon recent estimates. Fig. 9.6 is based on the authors' analysis of information included in the IPCC's October 2018 report and other reports. Numbers are approximate, reflecting the uncertainty in predicting the specific increase in CO_2 concentrations and the resulting increase in global temperature. The figure demonstrates what happens if actions are not taken to reduce greenhouse gas emissions. Increasing emissions lead to higher

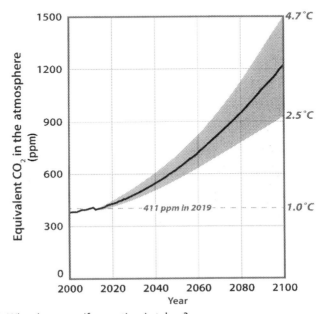

Figure 9.6 What happens if no action is taken?

concentrations of greenhouse gases in the atmosphere, shown as the horizontal lines in Fig. 9.6. The concentration in early 2019, 415 ppm, corresponds to a temperature increase of $1.0°C$.

The forecast of greenhouse gas emissions uses the same assumption as used in Chapter 14, Action Plan: Efficiency, Power, Transportation, and Land Use, which assumed that energy demand would increase about 1.25%/year with a corresponding increase in greenhouse gas concentrations. Recently, actual increases have been greater than 1.25%/year, so this forecast is conservative. Actual emissions could grow faster or slower depending upon the assumptions used and any actions taken to reduce emissions. This forecast does not assume any tipping points were reached that would trigger a faster growth in emissions.

- The U.S. government report also states that "over the next few decades (2021−50), annual average temperatures are expected to rise by about $1.4°C$ for the United States, relative to the recent past (average from 1976 to 2005), under all plausible future climate scenarios."[8] In addition, the report states that "without major reductions in emissions, the increase in annual average global temperature relative to preindustrial times could reach $5.0°C$ or more by the end of this century. With significant reductions in emissions, the increase in annual average global temperature could be limited to $2.0°C$ or less."
- There will be climate-related weather extremes such as storms, floods, and droughts. Forest fires would increase in frequency and intensity. The recent history of severe storms and floods in the southeastern United States and the increased wildfires in the western United States and Europe offer a preview.
- The sea level will continue to rise, causing more severe and more frequent flooding and higher storm surges. Ocean acidification will increase along with higher seawater temperatures. Coral would die off as it has in many places. Other sea creatures would be affected.
- With earlier spring snowmelts along with reduced snowpacks, water supplies in some areas would be reduced. Droughts will occur. Crop yields will decrease in many areas with crop failures in some. The deserts and tropics would expand.
- Due to poor agricultural, ranching, and wood harvesting practices, we are rapidly losing natural carbon sinks. In the Amazon, illegal logging and land theft are resulting in the destruction of tropical forests. In

[8] Wuebbles et al., *NCA4*.

fact, tropical forests have been so reduced due to logging, controlled burning, and forest fires that they are beginning to release more carbon than they sequester.[9] In 2019, the world was shocked to learn that vast regions of the Amazon rain forest were on fire. To create more land for cattle ranching and soybean farming, trees were felled, allowed to dry, and then set on fire. In August 2019 alone, there were more than 25,000 fires burning. Smoke spread more than 1000 miles south to blanket São Paulo.[10] Brazilian scientists estimate that humans have deforested roughly 16% of the entire Amazon basin. According to Carlos Nobre, a leading Brazilian climatologist, the tipping point for tropical forests is a loss of 20%−25% by deforestation. That loss is enough to convert much of the rain forest to savanna, releasing large amounts of CO_2 into the atmosphere as trees die and burn.[11] The loss of forests is accompanied by the extinction of some species.

- The pressure for some people to migrate would increase due to flooding, crop failures, or excessive heat and humidity.

The high cost of doing nothing

Regarding the U.S. president's assertion that "We cannot spend trillions on preventing climate change," studies confirm that the world is likely to spend more trillions—*if we do not take action.*[12,13,14] For example, perhaps as much as $60 trillion would be the cost of damage *just caused by the release of 50 Gt of methane.* Besides the damage due to sea level rise and coastal flooding of the U.S. coastal areas, the cost of disease and health impacts is impossible to predict. Much of the expense would be borne by

---

[9] Thomas E. Lovejoy and Carlos Nobre, "Amazon Tipping Point," *Science Advances* 4, no. 2 (2018): eaat2340, doi:10.1126/sciadv.aat2340.

[10] Julian Rosen, "Alarms from Amazon Fires," *Los Angeles Times*, August 26, 2019, p. A1. See also "On the Brink," *The Economist*, August 3, 2019, p. 14.

[11] Sam Eaton, "A Climate Tipping Point in the Amazon," *The Nation*, September 24/October 1, 2018, pp. 18−23.

[12] Julia Rosen, "Battling Climate Change Sooner, Not Later," *Los Angeles Times*, March 30, 2019, p. B2.

[13] Julia Rosen, "The Cost of Climate Change," *Los Angeles Times*, April 15, 2019, p. A2.

[14] Frank Ackerman and Elizabeth A. Stanton, *The Cost of Climate Change* (New York, NY: Natural Resources Defense Council, May 2008), https://www.nrdc.org/sites/default/files/cost.pdf.

developing countries experiencing extreme weather, flooding, droughts, destructive wildfires, and loss of crops.[15]

We cannot afford to "do nothing." We are already spending trillions. The cost of fighting worldwide wildfires in 2018, including property damage, is billions of dollars. Losses due to recent hurricanes and the disastrous typhoons in Japan, the Philippines, South China, and Hong Kong are billions more. Globally, the total runs into trillions of dollars. The lesson: pay now, or pay more later.

Doing nothing or not enough is clearly not acceptable. What needs to be done? Chapter 15, Can It Be Done?, covers this subject in more detail. Even if we do not see how we can do enough, fast enough, to solve the problem, we need to get started by taking actions that reduce the emissions of CO_2 and increase the earth's ability to absorb excess CO_{2eq}, that is, increase the sinks as well as reduce emissions. In parallel, we need to intensify development of CO_2 removal technologies and other coping mechanisms on the expectation that they will be needed.

---

[15] Nina Chestney, "Arctic Methane Release Could Cost the Economy $60 Trillion," *Reuters*, as noted in *Scientific American*, August 6, 2018, https://www.scientificamerican.com/article/arctic-methane-release-could-cost-60t/.

CHAPTER 10

Energy alternatives

While the combustion of fossil fuels is the main source of greenhouse gases, there are other significant sources from agriculture and land use changes, such as clearing the rain forest to create land to grow crops. This chapter will focus on the use of fossil fuels as raw materials and a source of energy and heat, and on their potential replacements. Fig. 10.1 shows world energy consumption by fuel for the past 28 years plus a forecast of future consumption if no additional changes are made to address global warming. Fig. 10.2 shows what segments of the economy use all this energy. About 28% is used as electricity.

If we are to phase out the use of fossil fuels, what are the alternatives? Some of the displacement can come from conservation and efficiency improvements. The rest has to come from substituting renewable energy sources; nuclear, solar, wind, biofuels (fuels derived from plants or animal wastes), and hydro.

There is no perfect solution. All the alternatives have weaknesses as well as strengths. For example, nuclear is an abundant source of carbon-free energy but has other problems that limit its potential. What are the advantages and disadvantages of alternative energy sources?

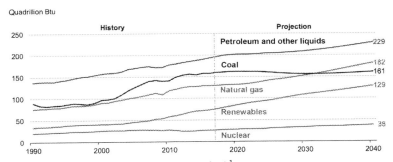

Figure 10.1 Global energy consumption by fuel.[1]

---

[1] Linda Capuano, U.S. Energy Information Administration, "International Energy Outlook 2018," Presentation for Center for Strategic and International Studies, Washington, DC, July 24, 2018, https://www.eia.gov/pressroom/presentations/capuano_07242018.pdf.

Reaching Net Zero.
DOI: https://doi.org/10.1016/B978-0-12-823366-5.00010-5

© 2020 Elsevier Inc.
All rights reserved.

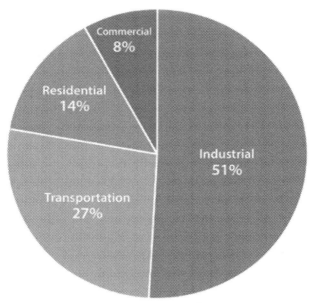

Figure 10.2 Global energy consumption by sector.[2]

The best way to keep fossil fuels in the ground is to have renewable forms of energy that are cheaper. They also have to be as available and dependable as the fossil fuels they replace. Solar, wind, and hydro energy have to become our main sources of energy, augmented by next-generation nuclear if radioactive waste management, siting, and other nuclear issues can be resolved. Solar and wind are abundant, increasingly inexpensive, and available to everyone and every country. Hydro is an important resource in the limited areas where it is available.

Renewables are becoming more cost-competitive with fossil fuels, especially due to improvements with solar cells, wind turbines, and batteries. The biggest barrier to substituting solar and wind power for fossil fuels is their intermittent nature, which makes it difficult to match electric supply with demand. This requires large utility-scale batteries or other means to store and release electricity as needed. This topic is discussed later in this chapter.

---

[2] Source: "International Energy Outlook 2019," p. 29, US EIA, Sept 24, 2019 https://www.eia.gov/outlooks/ieo/pdf/ieo2019.pdf

Fossil fuels: coal, oil, and natural gas

Fossil fuels are abundant and relatively inexpensive. We will not run out of fossil fuels for a very long time. To date, they have powered the industrialization of the global economy. We are still using obsolete 100-year old technology, burning fossil fuels, to provide most of the world's energy. Over the past 100 years or more, an extensive global infrastructure has developed to extract, process, transport, and make use of fossil fuels. This has to change, and will change, but change will not come easily.

When burning coal, oil, and natural gas to produce 1 million Btu of energy, different amounts of CO_2 are produced. Coal is the fossil fuel producing the most CO_2 per unit of energy produced while natural gas is the lowest. Table 10.2, later in this chapter, compares fossil fuels and bio-fuels. The use of fossil fuels has enjoyed a very substantial subsidy in most countries in that there is little or no cost associated with the discharge of CO_2 and pollutants into the atmosphere when burning fossil fuels to produce energy or heat. However, some countries are now imposing a carbon fee or CO_2 tax. Applying a reasonable cost to the discharge of CO_2 would be a major step toward reducing use of fossil fuels in favor of conservation and renewable sources of energy. Such a cost should not be thought of as a "carbon tax" as much as a charge for using the earth's atmosphere as a dump for greenhouse gases and other pollutants. As stated earlier, we have to always keep in mind that with greenhouse gases, mainly CO_2, what goes into the atmosphere stays in the atmosphere for a long time, increasing global warming.

Coal. Coal is a fairly inexpensive fuel and is abundant in many but not all countries. For example, Japan and Korea do not have significant coal deposits. Coal is still the main fossil fuel used for the production of electricity on a global basis. Using coal in China and India to meet a growing demand for electricity is a major reason CO_2 emissions have increased in the recent past.

Coal is carbon (C is the chemical symbol for carbon). When coal is burned to produce energy, all the energy is derived from converting carbon and oxygen into CO_2. In addition, coal includes many contaminants such as sulfur. During combustion, this sulfur is converted into oxides of sulfur, a major health hazard. Burning coal also produces particulates that can be released to the atmosphere. There is a wide variety of coal types available depending upon where it is mined. Some types have more impurities than others and produce more pollution in addition to CO_2.

If we want to reduce CO_2 emissions, reducing the use of coal should be our first priority where it is possible. Some countries do not currently have good options to reduce their use of coal without technical and financial support from the international community.

Oil and its derivatives. Oil and natural gas are hydrocarbons. That is, their molecules contain hydrogen atoms as well as carbon atoms. When hydrocarbons are burned to produce energy, a large percentage of the energy produced is from combining hydrogen and oxygen to form water (H_2O). This water leaves the combustion process as water vapor.

Oil is the source of essentially all of our liquid fuels such as gasoline, diesel oil, and aviation fuel. Oil is also the cheapest fuel to transport globally. There is a huge global infrastructure for oil transportation using ocean-going tankers and pipelines. The United States is the world's single largest consumer of oil (Photo 10.1).

At one time there was a concern that we were running out of inexpensive sources of oil and that oil prices would increase dramatically. However, new technologies for the exploration and extraction of oil have led to a global surplus of oil and lower oil prices. Fracking (oil and gas produced by injecting water under high pressure into subterranean

Photo 10.1 Offshore oil platform, Gulf of Mexico.

rock formations to create or open fissures to extract oil or gas) in the United States has had a major impact on the cost of producing oil and natural gas. It has also made it possible to tap oil deposits that were considered to be depleted or uneconomical to exploit. This has led to a surge in U.S. oil and gas production such that the United States is now tied with Russia and Saudi Arabia as the world's largest oil producers. U.S. oil production recently exceeded 12 million barrels/day (August 2019). At present the United States only imports about 20% of the oil used domestically, down from 60% about 10 years ago.

Oil prices are variable depending upon global supply and demand. Early in 2020 the price had been oscillating from \$US52 to \$US62/barrel, up from \$US30/barrel in early 2016. Oil prices were over \$US100/barrel from about 2009 (after the financial crisis) until 2014.

Natural gas. Natural gas, essentially methane (CH_4), is presently the cleanest fossil fuel, producing the lowest amount of CO_2 for a given energy output. The United States currently has an abundance of natural gas due in part to fracking. Natural gas is presently inexpensive compared to other fossil fuels, ranging from \$US2.00 to \$US3.00/MBtu pipeline cost (early 2020), or \$US7–\$US10/MBtu for commercial/residential customers. Because of the low cost of natural gas, it is displacing coal in the United States for electricity generation and has contributed to a decline in greenhouse gas emissions in the United States. However, methane is a potent greenhouse gas. Leakage of methane during the extraction and transportation of natural gas adds to greenhouse gas concentrations in the atmosphere. The demand for natural gas is increasing as a fuel and as a feedstock for chemical and fertilizer production. Natural gas is also the main fuel for home heating. This demand could eventually raise the cost of natural gas in the United States and other countries. Natural gas prices have been as high as \$US20.00/MBtu in the United States in the last 20 years.

The cheapest way to transport natural gas is by pipeline. Pipelines are expensive to build and require a right of way (property easement), which is often contested by property owners. The availability and capacity of pipelines can limit access to natural gas. Transporting natural gas across the ocean, for example, from the United States to Japan, is very expensive. Natural gas has to be converted to a liquid at a very low temperature and transported in expensive special-purpose ships. At the receiving end, it has to be converted back into a gas. This is an energy-intensive and expensive process that requires heavy capital investment and raises the cost of

Table 10.1 Fossil fuel types.

Fuel type	Advantages	Potential problems
Coal	The most abundant and lowest cost fossil fuel in many countries. Can be transported globally by rail and ship.	Produces more CO_2 per unit of energy produced than other fossil fuels. Is a major source of air pollution.
Oil	Abundant global supply. Global transportation network for low-cost distribution. Source of all liquid fuels: gasoline, diesel fuel, and aircraft fuel. Is a feedstock for plastics and other products.	Produces CO_2 and other greenhouse gases. A major source of urban air pollution. No readily available substitutes for liquid fuels. Oil price increases quickly impact global economy. Price can vary considerably.
Natural gas (methane)	Fossil fuel with lowest CO_2 emissions. Low levels of other pollutants. A feedstock for fertilizer and plastics production.	Produces CO_2. Methane is a greenhouse gas, leakage contributes to global warming. Requires pipelines for economical transportation. Price can vary considerably.

imported natural gas. This limits the potential for exporting natural gas compared to oil and even coal. Table 10.1 compares the fossil fuel alternatives.

Nuclear power

Nuclear power is an abundant source of carbon-free energy and is available 24/7. However, it has a number of disadvantages that make it an unlikely source of carbon-free energy in the United States and most other countries. Some countries such as China, Russia, and South Korea are expanding their nuclear power capacity. China and Russia are attempting to sell nuclear power plants to other countries, hoping to make this a major export business. Technical leadership for the design, construction,

Energy alternatives 129

and operation of nuclear power plants has moved to China. U.S. capabilities have declined.

The United States is the world's largest producer of electricity from nuclear power. Currently about 20% of United States electricity is generated by nuclear power plants. Despite early successes, for various reasons the United States has halted future attempts to use commercial nuclear power as a source of electricity. Other countries, such as France, continue to depend on nuclear power for electricity generation, but are not adding new nuclear plants to replace those that are nearing the end of their service life. In the United States, the 98 nuclear power plants in use are aging. Many are reaching the end of their service lives and will be shut down in the foreseeable future. Only one new nuclear power plant is still under construction in the United States and no other new plants are being planned. However, it is important that the United States continue to operate nuclear power plants for as long as possible. As they are eventually shut down, a large percentage of the power they produce could be replaced by coal- and gas-fired power plants, increasing CO_2 emissions.

In contrast to commercial nuclear power in the United States, the U.S. Navy has perfected the use of nuclear propulsion for submarines and aircraft carriers and has an outstanding record for the safe and reliable operation of high-performance nuclear power plants in demanding operating environments. In Chapter 13, Some Success and Failures, we discuss why commercial nuclear power in the United States failed and what needs to be done if commercial nuclear power is to be revived in the United States

Renewable energy

If we are to significantly reduce greenhouse gas emissions, we need to have a rapid transition to renewable forms of energy that do not produce greenhouse gases—mainly solar and wind energy. Biofuels have some potential. However, biofuels require fossil fuels for growing and processing plant material into useable fuel, essentially ethanol (alcohol). This detracts from their potential benefit. Biofuels also use land that could be used to grow food crops.

A major bottleneck preventing the widespread substitution of solar and wind energy for fossil fuels is their availability. Solar and wind energy output is dependent upon the time of day, the time of year, and the weather. Electricity generation peaks during the summer months and decreases during the winter months. Cloud cover decreases the output of solar cells. Currently, most electricity has to be consumed as it is produced due to the cost of storing it for future use. However, cost-effective means of storing large amounts of electricity for future use are being developed and the cost of utility-scale batteries is steadily declining.

An advantage of electricity is that it is easy, safe, and efficient to transport great distances by way of high-voltage transmission lines. It can be easily distributed to individual homes and businesses through local distribution networks.

However, there is a limit to the amount of intermittent power the electric grid can absorb. Some estimates say that an electric grid cannot include more than about 15% of solar or wind power generation due to its intermittent nature. In contrast, utilities require a high baseload of power generation that can only be provided by large central power stations—coal, natural gas, hydro, or nuclear. As more intermittent solar and wind power is added to the network, backup power units, usually gas turbines, are needed to meet demand when renewable sources of energy are not sufficient. There are also times when solar and wind produce more electricity than can be used by the network. Surplus power can be sold to customers outside the network provided that there is a demand for the power and a connection (see Fig. 10.3).

Solar energy. Extensive research and investment are being directed at means of storing electricity to better match the difference between solar and wind generation with demand. Improvements in energy storage and other means should facilitate the increased use of solar- and wind-produced electricity. Advanced batteries are being designed for a wide variety of applications, from tractors to buses.[3] Cost-effective batteries for electric grid operation are being developed.

In some locations, electricity from solar and wind generation is already cheaper than fossil fuels and nuclear. The trend is for these energy sources to become even less expensive and more competitive with fossil fuels. Fig. 10.4A and B shows the declining levelized cost (LCOE) of solar and wind energy without subsidies.[4]

---

[3] Anna Hirtenstein, et al., "Building Bigger Batteries," *Bloomberg Business Week*, September 17, 2018, pp. 26−27.

[4] Lazard, *Levelized Cost of Energy Analysis—Version 12.0*, November 2018, https://www.lazard.com/media/450784/lazards-levelized-cost-of-energy-version-120-vfinal.pdf. Used with permission from Judi Mackey, Managing Director, Global Communications, Lazard.

Energy alternatives

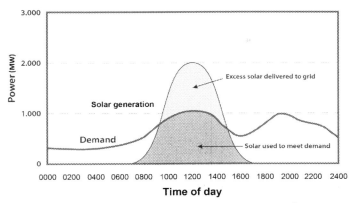

Figure 10.3 Mismatch between solar power supply and demand.[5]

Solar cells also allow the development of widely dispersed energy networks such as rooftop solar installations, allowing users to move away from large central power stations with transmission and distribution networks as the primary source of electric power. Wind energy can also contribute in some but not all locations. There are other forms of solar energy such as solar thermal generating stations. In this analysis, we focus on solar photovoltaic energy production ("solar cells").

The solar cell has been improved to the point that it is a viable solution for converting sunlight into electricity at a competitive cost. Investment in solar installations is increasing rapidly, which should further improve the technology and drive down costs. Solar cells are rapidly becoming more common as low-cost components to convert sunlight into useable electricity that is available for use in a variety of applications.

Wind power. Large wind turbine technology has advanced and continues to improve in cost and performance. There is considerable new investment in wind turbines, both in building new wind farms and in designing larger and more efficient wind turbines (see Fig. 10.5). Small wind turbines, used long ago in rural areas, are making a comeback but are expensive. In 2016, wind energy production exceeded hydroelectric energy production in the United States.

---

[5] This figure illustrates one of the challenges of solar energy. Power demand has two peaks, one roughly mid-day and a second in the evening. Solar energy can supply the mid-day peak, driving down the need for other sources. As the evening peak arrives, solar generation declines and the demand for backup power (in the absence of battery storage) surges. The resulting demand curve is sometimes referred to as a "duck curve," from its shape.

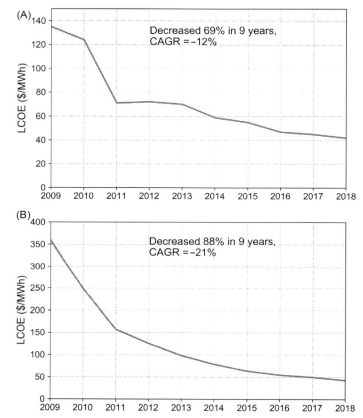

Figure 10.4 (A) LCOE Wind Power. (B) LCOE utility-scale PV solar.

Hydroelectricity. Hydroelectric power generation is considered to be a carbon-free source of energy. However, there is limited potential for adding to hydroelectric power generation. There are not many remaining dam sites to be developed and new dam projects are opposed by environmental groups. They object to damming rivers and flooding large areas needed to create reservoirs. Also, as global warming progresses, there may be droughts and other effects that could affect stream flows and reduce hydro potential.

In the U.S., there is a possibility that Hoover Dam may be unable to generate electricity at some time in the future. As the water levels fall (and it has been declining for the past decade), the power output of the dam's 17 turbines is derated from the nameplate value of 2074 MW down to a current value of 1592 MW. If the water level in Lake Mead

Energy alternatives 133

Figure 10.5 Trends in wind turbine design.[6]

drops another 100 ft. or so, the dam will no longer be able to generate electricity.[7] What is even more ominous is the possibility that decline of the water level in Lake Mead is irreversible. This could be another "Tragedy of the Commons" unless the seven states holding water rights can come to agreement on sensible reductions on upstream water use to allow the reservoir behind Hoover Dam to reach capacity.[8]

Biofuels. Biofuels (liquid or solid fuels made from plant materials or animal wastes) are at best carbon neutral, and then only if the carbon cycle is closed and the biological feedstock (the raw materials used to make biofuels) is replaced with new plant growth. CO_2 is emitted when biofuels are burned, the same as fossil fuels. An equivalent amount of CO_2 is reabsorbed over time if the biological feedstock is replaced with new growth that grows to maturity, but this takes time. If corn and sugar cane are used as feedstock, an equivalent amount of CO_2 can be absorbed

---

[6] Adopted from Jeremy Grantham, "The Race of our lives revisited," Exhibit 11, Giant Wind Turbines Illustrate the Speed of Change, *GMO*, August 2018, https://www.divestinvest.org/wp-content/uploads/2018/10/GMO.the-race-of-our-lives-revisited.pdf.

[7] Rod Kuckro, "Receding Lake Mead Poses Challenge to Hoover Dam's Power Output," *E&E News*, June 30, 2014, https://www.eenews.net/stories/1060002129.

[8] Michael Hiltzik, "Lake Mead Decline May be Irreversible," *Los Angeles Times*, February 10, 2019, p. C1.

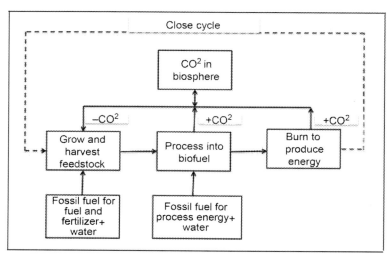

Figure 10.6 The biofuels life cycle.

by plant growth the next growing season. For wood pellets, it can take 25–50 years for the trees to grow to maturity (see Fig. 10.6).

In addition, biofuels need fossil fuels in the form of gasoline or diesel fuel for farm machinery, fertilizer, and electricity to grow, harvest, and process the feedstock into useable fuel. CO_2 from fossil fuel used to produce biofuel has to be added to the CO_2 released by burning the biofuel. For some biofuels such as corn-based ethanol, it can require up to 0.8 gallons of fossil fuel to produce one gallon of biofuel—a questionable benefit. Ethanol produced in Brazil using sugar cane is more productive and requires less fossil fuel to produce a gallon of ethanol.

With some biofuel feedstock such as corn, farmland is being diverted from growing food crops to fuel production. This raises food prices and could be a problem for countries that experience chronic food shortages.

A major offsetting disadvantage results if forests are cleared to produce farmland to grow biofuel feedstocks or produce wood pellets. Clearing forests eliminates a carbon sink by reducing the amount of plant life available to absorb CO_2 from the atmosphere. Reducing carbon sinks contributes to global warming by forcing more CO_2 into the atmosphere and oceans.

Currently used liquid biofuels, which include ethanol produced from crops containing sugar and starch and biodiesel from oilseeds, are referred to as first-generation biofuels. These fuels only use a portion of the energy potentially available in the biomass.

Table 10.2 Biofuels compared to fossil fuels energy content.

Fuel	Energy content	Amount of fuel for 1 MBtu	Kilograms of CO_2 produced per 1 MBtu
Wood chips	8–10 kBtu/pound	100–125 lbs	98–123
Coal	12.5 kBtu/pound	80 lbs	95
Biodiesel	16.3 kBtu/pound	61 lbs	80
Ethanol	10–11.5 kBtu/pound	87–99 lbs	75–86
Oil	130 kBtu/gallon (diesel fuel)	7.7 gallons	73
Natural gas	1 kBtu/ft.^{3}	1000 ft.^{3}	53

Biodiesel is produced, mainly in the European Union, by combining vegetable oil or animal fat with an alcohol. Biodiesel can be blended with traditional diesel fuel or burned in its pure form in diesel engines. Its energy content is somewhat less than that of diesel (88%–95%). Biodiesel can be derived from a wide range of oils, including rapeseed, soybean, palm, coconut, or jatropha oils. Because of the variety of feedstocks used, the resulting fuels can display a greater variety of physical properties than ethanol. Table 10.2 compares biofuel energy content with fossil fuels.

Biogas is created when organic material is digested by bacteria in the absence of oxygen, a process typical of sewage treatment plants, landfills, and rice paddies. During production, there is a solid byproduct in addition to the biogas. This can be burned as a biofuel or used as fertilizer. Biogas consists largely of methane. Biogas is created in landfills due to natural digestion of organic waste and can be collected for use as a fuel.

In less developed countries, dried animal manure is used as a biofuel for cooking and heating. Charcoal is also used for these purposes. Charcoal is produced by heating wood in the absence of oxygen to remove water and other constituents to produce a black carbon fuel that burns much hotter than wood. Unfortunately, the production of charcoal can deplete woodlands and forests that are natural carbon sinks.

Some next-generation biofuels are being developed. They would use switchgrass, algae, and other plant materials as a feedstock. Next-generation biofuels could be produced from agricultural and forestry waste. Unlike first-generation biofuels, these fuels could significantly reduce CO_2 emissions, and in addition they do not use food products as the feedstock. Some types have even enabled better engine performance. However, progress has been slow in producing useable fuels from waste plant material, algae, and

Table 10.3 Renewable fuel alternatives.

Energy form	Advantages	Potential problems
Solar (PV)	Fuel-free, universally available. Cost of solar produced electricity is competitive and decreasing. Electricity easily distributed via the electric grid.	Intermittent output, varies by time of day, time of year, and with the weather. Requires large land area. Not practical at certain latitudes.
Wind	Fuel-free, available in many locations. Cost of wind-produced electricity is competitive and decreasing. Electricity easily distributed via the electric grid.	Not universally available. Intermittent output, varies by time of day, and with the weather. Requires large land area, hazard to birds.
Hydrogen	Powers efficient fuel cells without producing greenhouse gases.	Limited supply. May never be cost-competitive. Only a green fuel if produced using electricity from renewable or nuclear sources.
Biofuels	Can be substituted for gasoline and diesel fuel.	Requires large land area to grow feedstocks. Needs fossil fuel inputs. Competes with food crops for land and other resources. Not as green as solar and wind. If forests cleared to produce farmland for biofuels, benefits will be negated. Produces farm waste.
Hydro	Dependable source of electricity.	Limited availability. Requires dams and large reservoirs.

other alternatives.[9] As shown in Fig. 10.6, all biofuels are at best carbon neutral and then only if their plant feedstocks are replaced by new growth that reabsorbs the CO_2 produced when the biofuel is consumed.

Also, it is important to recognize that some end uses cannot be replaced by renewables. For some processes, such as the production of plastics and fertilizers, it may not be practical to replace oil and natural gas.

Table 10.3 lists the primary sources of renewable energy along with their advantages and disadvantages.

---

[9] Brian Barth, "The Next Generation of Biofuels Could Come from These Five Crops," *Smithsonian Magazine*, October 3, 2017, https://www.smithsonianmag.com/innovation/next-generation-biofuels-could-come-from-these-five-crops-180965099/.

Hydrogen. Hydrogen can be a green fuel in that its combustion yields water (H_2O), not CO_2, so long as it is produced by electricity from renewable sources. Hydrogen-powered fuel cells are more efficient than gasoline or diesel combustion engines. Hydrogen fuel cells are a proven technology and prototypes of hydrogen-powered vehicles are on the road today.

However, hydrogen is not a source of energy but an energy carrier similar to electricity. Hydrogen is very reactive, and there is no free hydrogen—no hydrogen mines or wells. There are several ways to produce hydrogen to be used as a fuel. Hydrogen can be produced by electrolysis, which separates water into hydrogen and oxygen using electricity. Another method is by natural gas reforming, which produces hydrogen from natural gas. This process uses fossil fuels to produce hydrogen and is not green energy.

If solar and wind power can produce abundant, low-cost electricity, it may be possible to use renewable electricity to produce hydrogen at a price that would make it a potential fuel source. Hydrogen is a gas at room temperature. It has to be compressed at high pressures to reduce its volume before it can be used as a vehicle fuel.

Converting CO_2 into usable liquid fuels. It is unlikely that all uses of liquid fuels can be eliminated. Examples are for aviation, ocean shipping, and some mobile or portable power uses such as emergency generators. As discussed earlier, biofuels may be an alternative source of liquid fuels to replace gasoline, diesel fuel, and aviation fuel. However, biofuels use agricultural land that could be used for food production.

An alternative is to convert CO_2 removed from the atmosphere to produce these liquid fuel alternatives. It has been demonstrated that fuels that are drop-in replacements for gasoline, diesel, and aviation fuel can be produced. The challenge is to produce these fuels in large quantities at competitive prices. Like biofuels, synthetic fuels do not actually remove CO_2 from the atmosphere, since the process recycles the CO_2 into fuel that will release CO_2 back into the atmosphere when it is consumed. The benefit is that it avoids adding *additional* CO_2 if fuels produced from oil were used. This only works if the entire process is powered by electricity from renewable sources.

A carbon fee would help by raising the cost of fossil fuels and make synthetic fuels more competitive. Using today's technology, a carbon fee of \$US200/ton of CO_2 or more would be required for synthetic fuels to be price competitive. A carbon fee of this amount would be impractical to apply suddenly. A lower fee would be appropriate to start with, followed by annual increases until the fee is equal to or greater than the cost of removing the CO_2 and offsetting the damage done by any additional air pollution or global warming.

The process is being pioneered by *Carbon Engineering* of Canada and others.[10] Their goal is to make fuels by removing the carbon from the CO_2 and combining it with hydrogen produced by electrolysis to produce a gas or liquid hydrocarbon fuel. This could be a synthetic natural gas or a gasoline or aviation fuel substitute. This process is energy-intensive and only beneficial if renewable energy is used to power the process.

Airlines need alternatives to jet fuel that do not add to the weight of the aircraft or take up more volume. Aircraft account for only about 2.0% of the world's fossil fuel use, but airline travel is growing steadily and increasing its associated greenhouse gas emissions. Airlines mainly use kerosene as jet fuel. Most of this fuel is used by commercial aircraft with a small percentage being used by military aircraft. Airlines release carbon dioxide, nitrous oxides, water vapor, sulfates, and soot at high altitudes, which magnifies their impact.

For airlines, fuel accounts for about 25% of their operating costs, second only to personnel costs. Fuel costs are variable and have accounted for as much as 32% of airline costs recently. This gives the airlines an incentive to improve efficiency to reduce fuel costs. Although it is technically feasible to develop renewable alternatives to jet fuel, it is likely that these alternatives will cost more. Using them would increase the cost of airline travel.

Energy storage. Energy storage is a major issue that needs to be addressed as part of efforts to reduce global warming. Electricity storage is an essential component of electricity distribution systems to make the use of variable renewable energy practical. It allows electricity suppliers to match the supply from renewable energy sources, solar and wind, with electricity demand, which is also variable. Batteries can be used to store excess electricity when supply exceeds demand, and supply electricity to the grid when demand exceeds supply.

There are several main applications for energy storage. First are batteries for mobile applications. This can be further divided into batteries for handheld and other small devices or for electric vehicles that need much more power. The second is large utility-scale energy storage systems. These are fixed installations with very large capacities.

---

[10] See Carbon Engineering, https://carbonengineering.com/.

Energy alternatives

The biggest difference between these two broad applications is that mobile applications from vehicles to laptops require batteries that are lightweight and compact. Electricity storage for nonmobile applications can be much larger and heavier to provide more storage capacity.

Battery technology today is recognized as the main problem limiting the range of electric vehicles. Batteries are also the most expensive component of most electric vehicles and largely determine the price of the vehicle. The electrolyte, an essential component of lithium-ion batteries, incorporates a flammable liquid. Failure of a lithium-ion battery can result in a fire as has been reported in the press from time to time.

Considerable research and development money is being spent on improving the cost, capacity, and safety of batteries. Improved performance, mainly faster charging times, is also an objective.

Because of its lightweight, the preferred means of storing electricity for mobile applications is the lithium-ion battery. It is used in everything from laptops to automobiles. Lithium is the lightest metal in the periodic table and has the greatest electrochemical potential for electricity storage by weight. It is the preferred choice for mobile applications today. The increasing production of lithium-ion batteries for vehicles and other electricity storage applications is reducing the cost of these batteries and expanding their use. A recent Bloomberg News report states that the cost of energy from utility-scale lithium-ion batteries has fallen 35% since early 2018, a dramatic drop, to an estimated $US187/MWh.[11] The cost of lithium-ion battery storage is becoming increasingly more competitive.

Lithium-ion batteries are also used in large, utility-scale, energy storage systems for electric utilities. According to the U.S. Energy Information Administration, over 80% of large-scale battery storage today is based upon lithium-ion battery technology. However, other battery designs, such as flow batteries, are being developed for large, utility-scale, energy storage. Since size and weight are not a limitation for nonmobile batteries, other storage technologies might be superior to lithium-ion batteries for fixed installations.

The largest solar electric generation station in the United States currently is the Solar Star power plant in California, rated at 579 MW. Plants as large as 1000 MW are operating in China. A giant 1600 MW solar plant has recently been announced in Egypt. In April 2019, renewable

[11] H. J. Mai, "Electricity Costs from Battery Storage Down 76% Since 2012," *Bloomberg New Energy Finance*, March 26, 2019, https://www.utilitydive.com/news/electricity-costs-from-battery-storage-down-76-since-2012-bnef/551337/.

sources in the United States generated more electricity than coal for the first time.[12] The largest coal-fired plants in the United States are being closed as a result of no longer being economical.

Coal-fired generating capacity in the United States (2018) was 246,000 MW and dropping rapidly as plants were being retired. However, to effectively replace coal with solar-generated electricity, it is necessary to have a means to store electricity in order to serve nighttime loads, typically about one-third of the daytime load. To supply this load would require:

$$(246,000 \text{ MW}) \times 0.33 \times 12 \text{ hours} = 974 \times 10^3 \text{ MWh}$$

Current U.S. electricity storage capacity (large systems) is 31,200 MW (2019), over 90% of which is pumped hydro. Battery storage is currently 900 MW and is expected to triple by 2023.[13] To enable solar energy to replace coal-generated electricity, U.S. storage capacity would have to be increased. In 2017, Tesla completed the world's largest lithium-ion battery storage facility in Australia, rated at 100 MW power capacity and 129 MWh storage capacity. It would require about 7600 such plants to enable solar power to replace 100% of U.S. coal-generated electricity.

Various energy storage systems are being considered, including lead-acid, lithium-ion, sodium-sulfur, and "flow" batteries (vanadium redox or zinc bromide). For larger storage capacity, either compressed air energy storage (CAES) or pumped hydro has been considered.[14]

For power storage (to meet short-term power demand), plants such as the Tesla's Australia installation described above would need about 350 m^3 for storage. This is a cube, roughly 7 m (23 ft.) on a side. Some of the other storage concepts would require 10 times as much space.

Internationally, utility energy storage is projected to double six times by 2030, involving a capital investment of over $US100 billion. This would increase storage capacity from 56 to 300 GWh and 125,000 MW of capacity, on par with 128,000 MW of pumped hydro capacity world-wide. In addition to the United States, most of this growth will occur in seven countries: China, Japan, India, Germany, United Kingdom,

---

[12] Energy Information Administration, "U.S. Electricity Generation from Renewables Surpassed Coal in April," *Today in Energy*, June 26, 2019, https://www.eia.gov/todayinenergy/detail.php?id=39992.

[13] University of Michigan Center for Sustainable Systems, "U.S. Grid Energy Storage Factsheet," 2019, http://css.umich.edu/factsheets/us-grid-energy-storage-factsheet.

[14] APS Panel on Public Affairs, *Integrating Renewable Electricity on the Grid* (Washington, DC: APS Physics, no date), https://www.aps.org/policy/reports/popa-reports/upload/integratingelec.pdf.

Australia, and South Korea. As a demand increases, costs are projected to decline from \$US700 to \$US300/kWh by 2030.[15]

Portland General Electric is making plans to build the country's first large-scale energy facility that combines wind turbines, solar panels, and battery storage in Eastern Oregon. This will enable the utility to generate 50% of its electricity from renewable energy and replace capacity of a coal-fired plant that is being retired.[16]

In Texas, a solar energy project that includes the world's largest battery combined with a large solar farm is under development to provide electricity to the oil fields. The expansion of oil production and the higher energy needs of fracking, a new oil drilling technique, are increasing the need for electricity to power pumps and other equipment in the oil field. In Utah, a 1-MW energy storage system has been announced. This is to store enough electricity to support 150,000 homes. It is a hybrid system that includes several technologies. It will store some energy as hydrogen generated using electricity from a solar farm. This hydrogen will be used to power a gas turbine to produce electricity when needed. Compressed gas will be stored in salt caverns for use to power generators to produce electricity. In addition, there will be a large flow battery for electricity storage. Finally, there will be fuel cells that are powered by hydrogen to produce electricity as needed. This complex system for energy storage should be an important demonstration project for utility-scale electricity storage.

Actually, the largest energy storage systems in use today do not use batteries. Pumped hydro accounts for an estimated 95% of electricity storage based upon storage capacity. Pumped hydro works by using electric pumps to pump water to a dam at a higher elevation when electricity supply exceeds demand. When demand exceeds supply, the stored water is allowed to flow through turbines at a lower elevation to produce electricity needed by the electric grid. This system works well and is very reliable. The United States currently has about 20,000 MW of pumped hydro storage capacity. However, there are a limited number of sites that can be used for pumped storage and most have already been developed.

---

[15] Jeff St. John, "Global Energy Storage to Double 6 Times by 2030, Matching Solar's Spectacular Rise," *Bloomberg New Energy Finance*, as quoted by GTM, November 21, 2017, https://www.greentechmedia.com/articles/read/global-energy-storage-double-six-times-by-2030-matching-solar-spectacular.

[16] Courtney Flatt, "Portland General Electric Set to Build 1st-of-its-Kind Renewable Energy Site," *OPB*, February 16, 2019, https://www.opb.org/news/article/eastern-oregon-solar-wind-battery-renewable-portland-general-electric/.

There is also public opposition to the construction of additional dams to expand pumped hydro capacity. To the extent that pumping utilizes fossil fuel—generated electricity, pumped storage does not provide any reduction in CO_2 emissions.

There are other means of storing electricity, but these have limited potential compared to battery storage and pumped hydro. They include flywheels, compressed air energy storage systems, and thermal systems that store heat in molten salts.

Smart grids. In 1879 Thomas Edison illuminated several buildings with his new electric lamp. The buildings were connected with a direct current (DC) distribution system that brought power from a generator. In 1886 George Westinghouse was promoting a different approach—using alternating current (AC), which he claimed had several advantages. Most importantly, transformers could be used to increase the voltage and transmit power over longer distances—a feature not possible with DC. As we know, Westinghouse's system won the day, and the majority of the transmission and distribution systems in use today are AC, although DC is used when large blocks of power are transmitted at high voltages over long distances.

Beginning in 1890, the electric grid in the United States has been expanded and improved over the decades. Today it includes more than 9000 generating plants, with over 1,000,000 MW of generating capacity, connected to 300,000 miles of transmission lines spanning the country.[17] In many regards it is an engineering marvel. Interconnections between utilities enable power to be shared or shifted if one supplier has an equipment failure. Similar developments have occurred in Europe, Asia, and the balance of the world, although there are still many regions that are not interconnected or are lacking a reliable source of electricity.

The basic grid has several components. The first is a generating station that supplies electricity to the transmission system. At the transmission system, the voltage is "stepped up" (increased by a transformer) to reduce the capital cost of the transmission lines. The transmission system conveys power from the generating station to the vicinity of customers. There the voltage is reduced in a distribution station for delivery to factories and homes in the local area. There are controls and sensors to detect the "demand" or requirements for more power. As more users require

---

[17] "What is the Smart Grid?," SmartGrid.gov, U.S. Department of Energy, accessed February 9, 2020, https://www.smartgrid.gov/the_smart_grid/index.html.

electricity, a utility brings additional generation capacity online to match the load.

Except for some refinements, this basic system has not changed in 100 years. It is a "one-way" system, sending power to users who are connected to the distribution system. At present, large quantities of electricity cannot be stored for later use. When it is necessary to match supply with demand, it is done by adjusting the supply. To do this, utilities use large, baseload power plants such as nuclear or coal-fired power plants. These supply most of the load. Peaking plants, such as gas turbines that can increase or decrease their electrical output quickly, are used to match changes in demand. Changes in demand can also be met by buying or selling electricity with other utilities.

Today, the traditional electric grid faces a number of challenges. It is not set up to handle "two-way" flow of power. Suppose you are a residential customer with a rooftop solar system. During daylight hours you may produce more electricity than you consume, so your excess power needs to be sent back into the grid. But at night, you need to draw power from the grid. As a minimum you require a special electric meter that will record two-way flow of current. In addition, utilities need to add improved controls at their distribution stations to manage two-way flows of electricity. As utilities start adding large blocks of renewable energy to the system, the problem becomes more complex. Solar power is only available at best 10 hours per day. Even then it is not constant, as the amount of daylight is influenced by the seasons (winter vs summer) and by the weather (sunny vs cloudy or overcast days). Wind power is likewise subject to variations. The "one-way" grid has expanded to the point where sometimes weather or equipment failures cannot be handled and blackouts can occur. These problems will be augmented by the addition of distributed generation sources.

There is also an issue of economics. Utilities want to use their most economical sources of generation at a given point in time. The most expensive sources are peaking units. Utilities have different rate schedules depending on the type of customer and its demand for power. One rate schedule is called "time-of-use" and has higher rates when the customers need power at peak periods.

Utilities hope to respond to these new complexities by building "smart grids." This will be accomplished by employing digital technology that allows two-way communication between the utility and its customers. The smart grid will be based upon changes to transmission lines to make their operation more automated. The system will include digital controls, computers, and new sensing technologies to enable the grid to respond quickly to changes in electrical demand.[18]

When fully realized, the smart grid offers a number of benefits to utilities and users alike. They include:

- more efficient transmission of electricity,
- quicker restoration of electricity after an outage,
- reduced operational costs for utilities and ultimately lower costs for consumers,
- reduced peak demand,
- ability to integrate large-scale renewable energy systems, and
- improved security.

Proponents believe that a smart grid will also add resiliency to electric power distribution systems, making it better prepared to handle emergencies such as severe storms, earthquakes, large solar flares, and terrorist attacks. It will allow automatic rerouting when equipment fails or outages occur.[19]

Utilities are approaching smart grid development in various ways—some in a conservative manner, others more aggressively. Modifying the grid involves expense, regulatory approvals, and customer acceptance. Several steps are necessary. One involves retrofitting customers with "smart meters." These are meters that sense the customer's demand for power and transmit it to the utility so it can optimize its generation and distribution of power. A further refinement is to allow the smart meter to also communicate with the customer, so he or she can understand power demand at any point in time and make decisions. For example, certain operations such as using a clothes dryer could be done at night when the rates are low, rather than during peak times. Smart grids with smart meters will also have some flexibility to adjust the load as well by shedding non-essential loads such as clothes dryers when demand for electricity exceeds supply.

For many utilities the first step is automating the transmission and distribution system. This involves enhanced equipment monitoring and analysis capabilities within substations to increase reliability. New techniques improve the ability to locate faults and respond more quickly with intelligent relays and substation local area networks. Similar developments

[18] Ibid. p. 1.
[19] Ibid.

Energy alternatives   145

Photo 10.2 Micro grid, Borrego Springs, California.

would be undertaken in the distribution system. While the requirements for transmission systems are different from those used in the distribution system, the basic concept is the same. Improved sensor and communications capability is considered the key to automation[20] (Photo 10.2).

The traditional electricity grid is not designed to accommodate distributed energy sources such as wind and solar generators. Examples of the types of problems encountered include frequency and voltage variation, inability to respond quickly to load changes, and reduction of transmission system reliability. Smart grids will be better able to handle variable supply. This will be important for California, Arizona, Hawaii, and countries such as Italy, Spain, Japan, and Australia that are expanding the use of renewable sources of energy.

The micro grid. The concept of the micro grid is to create an integrated utility system consisting of electrical generation capability, energy storage for a predetermined time, and a distribution system to a local network of consumers. In one form or another, such small, unified systems have existed for some time, for example, serving a university campus or

---

[20] Scott Madden Management Consultants, "Smart Grid Planning: The Business Case for Transmission and Distribution Automation," February 2011, p. 7, https://www.scottmadden.com/wp-content/uploads/userFiles/misc/4d4e971708fab5e1c83d71b411345483.pdf.

medical complex. One of the principal characteristics of such systems is their ability to operate independently for short periods of time in the event of a general power failure. Micro grids are also being considered as a means of reducing damage from wildfires.

Partly in response to California's deadly Camp Fire that killed 86 people and destroyed the town of Paradise in 2018, regulators are giving new emphasis to the deployment of micro grids. In the aftermath of the fire, it has been determined that the fire was started by a Pacific Gas and Electric (PG&E) transmission line. When a series of fires broke out in northern California, PG&E considered shutting down the system but then delayed the shutdown to prevent customer outages. Had there been a micro grid system in the town of Paradise, it is possible that the utility system could have been disconnected by shutting down the transmission line to the Paradise area when the fire danger was high. With this in mind, the California Energy Commission has recently awarded $US50 million in grants to micro grid demonstration projects.

California is home to an advanced micro grid system developed by San Diego Gas and Electric Company. It is located in Borrego Springs, California, a remote desert community of 3400 residents, located about 145 km (90 miles) northeast of San Diego. The city is served by a single transmission line that traverses the mountains separating the Anza-Borrego Desert from the coastal plain. In the past there have been outages that left the city without power for days. The Borrego Springs micro grid has the capability of powering the city for several days in the event fire or equipment failure causes the transmission line to be taken out of service. In addition, the Borrego Springs system has been designed as a micro grid demonstration project.

Borrego Springs has two solar farms, one located near the municipal airport and a second private one operated by Clearway Energy Corporation, rated at 26 MW. These solar farms, along with 700 kW of rooftop solar capacity from private residences, convey power to the micro grid equipment yard. The yard contains switchgear, two 1.8-MW diesel generators, a large Saft 1.5-MWh lithium–ion battery, three smaller Saft 50-kWh lithium–ion batteries, and a 12-kV ultra-capacitor bank. The lithium–ion batteries store electrical energy for use during nighttime operation. The diesel generators provide backup power in the event of extended outages or failure of other equipment. The ultra-capacitor serves to smooth out short-term shifts in demand. In the event of a failure of the incoming transmission line, the micro grid switches over to independent operation, a process known as "Islanding."

The micro grid system has the potential to provide an independent electrical power system using renewable energy. It would be particularly advantageous in remote areas where significant cost savings could be obtained by avoiding the necessity of building expensive transmission lines.

Competitors to traditional utilities. Peter Drucker, a renowned management consultant, is credited with saying "innovate or die." This principle could apply to electric utility companies that fail to respond to changing energy economics quickly enough or that refuse to listen to their customers' desires for renewable energy. In the past, some cities in California revolted against their utilities and formed their own utility districts, examples being Sacramento, Burbank, and Anaheim. Now new competitors have emerged. Community Choice Aggregators, or CCAs, do not own generating plants nor operate their own grid. Instead, they buy and sell electricity that flows through utility grids. They are signing long-term deals for new renewable energy generating capacity, which is spurring growth.[21] In California, *The Clean Power Alliance* is becoming the new electric utility for many cities. In cities that have opted for the Clean Power Alliance, customers are automatically switched from their traditional utility. It offers three rate plans, one that is cheaper than the traditional utility, one that is on par with the utility, but is based on 50% renewables, and one that is based on 100% renewables and is 9% more expensive. It is possible to opt out and return to the traditional utility, but so, far few customers have done so. Clean Power Alliance now has 1 million customers and is investing in its own wind turbines.[22]

---

[21] Sammy Roth, "State's Changing Energy Landscape: New Entities Offer Clean Alternatives and Take Customers from Utilities," *Los Angeles Times*, January 27, 2019, p. C2.

[22] Sammy Roth, "Southland Is Plugging into a New Utility: Clean Power Alliance Is Becoming Many Cities' New Electric Utility. Here's What It Means for Residents," *Los Angeles Times*, February 1, 2019, p. C1.

CHAPTER 11

Unique problems of major contributors to global warming

There are no easy solutions. Each country has major practical problems to overcome to reduce greenhouse gas emissions.

The six largest greenhouse gas emitters are China, the United States, India, Russia, Japan, and Germany, accounting for about two-thirds of global greenhouse gas emissions (see Fig. 6.2). At a minimum, these countries have to work together to solve this problem. However, each of these countries has unique problems that they have to deal with.

What can we learn from Germany?

Germany is the world's fourth largest economy and made a national commitment to address global warming about 20 years ago, ahead of the United States and other major developed countries.

In 2000, Germany passed a major green initiative that forced electricity providers to purchase renewable energy at fixed prices for 20 years to guarantee private investors a return on their renewable energy investments. This guarantee led to huge private investments in solar and wind energy across Germany.

In 2007, Germany committed to cut greenhouse gas emissions from electricity production 40% from 1990 levels by 2020 and by 95% by 2050. This was referred to as the Energiewende meaning "energy turn-around" or "energy transformation." These goals were more ambitious than those set by the European Union, a 20% reduction in greenhouse gas emissions by 2020 and 40% by 2050.

Germany rapidly achieved a 28% reduction in greenhouse gas emissions from 1990 when East and West Germany were reunited, to 2009, the year following the 2008 financial crisis. Part of this rapid reduction was due to shutting down some dirty coal-fired power plants and closing energy-inefficient factories in East Germany. In addition, some

Reaching Net Zero.
DOI: https://doi.org/10.1016/B978-0-12-823366-5.00011-7

© 2020 Elsevier Inc.
All rights reserved.

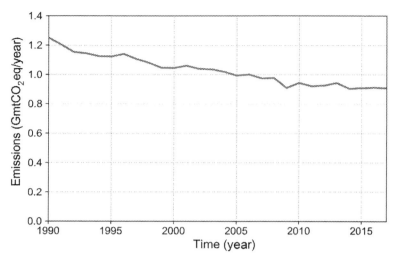

Figure 11.1 Germany's greenhouse gas emissions compared to 1990, in $GmtCO_{2eq}$.[1]

energy-intensive production, such as steelmaking, was exported to China. A major change was the greater use of wind and solar energy to produce electricity.

In spite of this early success, Germany's total greenhouse gas emissions have been flat for over 10 years at about 910 $MmtCO_{2eq}$/year (see Fig. 11.1). There was a reduction in 2018 to 866 $MmtCO_2eq$/year.

In 2018, the German government decided to give up on the country's goal for 2020. Cutting CO_2 emissions by 40% by 2020 was clearly unrealistic.

The largest source of Germany's greenhouse gas reduction was due to the increased production of electricity using solar and wind. The transportation sector reduced its emissions by only 1% since 1990, highlighting the difficulty of reducing or replacing the use of gasoline and diesel fuel for transportation. Any significant reduction for transportation will probably have to wait for the increased use of electric vehicles and perhaps hydrogen-powered vehicles (Table 11.1).

A key component of Germany's effort to increase the use of renewable energy for electricity production is the requirement that transmission system operators purchase solar energy generated by both small- and utility-scale solar installations at a 20-year fixed price referred to as the "feed-in

---

[1] German Environment Agency, "Indicator: Greenhouse Gas Emissions," Umwelt Bundesamt, April 25, 2019, https://www.umweltbundesamt.de/en/indicator-greenhouse-gas-emissions.

Table 11.1 Germany's greenhouse gas reductions.

Sector of the economy	Annual emissions (Mt CO_2 equivalent)		
	1990	2018	Change (%)
Electricity production	466	311	− 33
Buildings	210	117	− 44
Transportation	163	162	− 01
Industry	284	196	− 31
Agriculture	90	70	− 22
Other	38	10	− 74
Total	1251	866	− 31

tariff" (FIT). The operators in turn sell the power to the wholesale market at competitive prices, which fluctuate but are invariably lower than the FITs paid to renewable energy producers. The difference is not subsidized by the government. The difference is paid by Germany's electricity customers in the form of a renewable surcharge that appears on electric bills. Some industries are exempt from the renewable surcharge so that they remain competitive with foreign companies. This leaves households and small businesses to shoulder more of the burden. This increases their surcharge about 30%.

An effect of these well-intentioned policies that guarantee high rates for electricity from renewable sources is that Germany has the highest electricity rates in Europe, tied with Denmark, another country committed to renewable energy. The average cost of electricity in Germany is about 29.5 eurocents/kWh. This equals 33.5 U.S. cents/kWh compared to a much lower U.S. average of 12 U.S. cents/kWh.

Germany also needs to reduce its dependence on coal-fired power plants. Germany is Europe's largest producer and user of coal, which accounted for 35% of net electricity production in 2018. While Germany has made progress in using renewable sources of energy, the country has a strong coal lobby. That lobby's influence coupled with an initial shortfall in energy generation due to Germany's 2011 decision to phase out nuclear power has left the country stuck with fossil fuels longer than it intended.

Germany uses large quantities of natural gas and gets over 50% of its gas from Russia. They are in the process of building a major pipeline, Nord Stream 2, which will increase Germany's dependence on gas from Russia.

Germany also uses wood pellets from the United States for electricity production and counts wood pellets as a biofuel. However, producing electricity from wood chips or pellets emits more CO_2 and other pollutants than using coal. The net benefit only comes if the trees used to produce wood pellets are replanted and absorb the CO_2 emitted as the tree grows to maturity in 25—50 years or more.

In the aftermath of the Fukushima nuclear disaster in Japan, Germany shut down nine nuclear power plants to date and the eight remaining plants will be shut down in the future. Germany will have to replace the energy produced by these nuclear plants with renewable energy. If a portion of this energy is replaced using fossil fuels, then CO_2 emissions would increase.

More recently, Germany announced a bold decision to close all 84 coal-fired power plants by 2038. These plants account for 35% of Germany's electricity generation. To accomplish this, the government plans on spending $45 billion to mitigate the effect of the closures on the economies in the coal regions. With this measure, Germany believes that it can meet its commitment of a 95% reduction from 1990 CO_2 levels by 2050. [2]

The public initially strongly supported Germany's Energiewende program but are now concerned about the cost and limited progress in meeting Germany's reduction targets. Germany is having trouble with public opposition to additional wind turbines and transmission lines needed to increase its use of renewables. This public opposition has caused some transmission lines to be buried. This is much more expensive than carrying transmission lines on towers but was necessary to placate public opposition.

Germany only has coal as a fossil fuel. They import essentially all their natural gas and oil from Russia and the Middle East. These energy sources are thousands of miles away. In pursuing renewables, solar and wind, they are trying to satisfy their energy needs domestically. This is bound to fail. Germany probably does not have the potential to meet all its energy needs with locally produced renewables no matter what the cost. Germany's solar and wind potential is limited by its climate and northern location. To meet its goals for renewable energy, Germany should probably import a high percentage of its renewable energy similar to its fossil

[2] Erik, Kirschbaum, "Germany to Close Coal Plants by 2038," *Los Angeles Times*, January 27, 2019, p. A3.

fuel imports. Why not import renewable energy in the form of electricity from distant places such as the Sahara in North Africa or Kazakhstan with its huge deserts? The world's deserts and windy plains should become the new Saudi Arabia.

The United States fails to take a leadership position

The United States recently became the world's largest oil producer, slightly ahead of Saudi Arabia and Russia. The United States is also becoming energy-independent in that it will soon produce as much fossil fuel energy as it consumes. This is due to the fracking revolution that allows U.S. oil companies to economically extract more oil and gas from existing oilfields. By 2019, U.S. oil production had increased about 130% over the past 10 years, and exceeds the last peak in domestic oil production that occurred in 1983, over 30 years ago. The Department of Energy forecasts that the United States will be a net exporter of oil by the end of 2020.[3] The United States also has large reserves of oil, natural gas, and coal.

In sharp contrast to its significance as a fossil fuel superpower, the United States is in the process of withdrawing from the Paris Agreement and does not plan to do much if anything at the federal level to deal with global warming. When President Trump announced his decision to withdraw from the Paris Agreement, he claimed that it was "an agreement that disadvantages the United States to the exclusive benefit of other countries." Most important, it would have required the United States to hobble a major source of income: its vibrant fossil fuel industries.

So far, the United States has not formally withdrawn from the Paris Agreement and continues to participate in international meetings. By the rules of the Paris Agreement, the process of withdrawal can only begin after 2020, soon after the next presidential election. There is a chance that the United States will eventually reenter the agreement. However, the United States government is not an active participant and is not showing any leadership in solving this global problem. Despite the government's

[3] See U.S. Energy Information Administration, https://www.eia.gov/; The GlobalEconomy.com, https://www.theglobaleconomy.com/; and Central Intelligence Agency, "The World Factbook," site last updated February 5, 2020, https://www.cia.gov/library/publications/the-world-factbook/geos/xx.html.

position, many states, cities, and some private businesses say they will abide by the Paris Agreement commitments.

Increased use of solar and wind energy for electricity production in the United States has contributed to reducing greenhouse gases. The biggest contributor, however, was the ongoing substitution of natural gas for coal for electricity production. This reduction was accomplished without any government mandates or subsidies. It was the result of the energy market responding to the relative cost of gas compared to coal. Due to fracking, there has been an increase in natural gas supplies and lower prices relative to coal. Since 2010, about 40% of coal plants have been shut down or designated for closure. An estimated 88% of coal power plants were completed before 1990 and are reaching the end of their service lives. It is unlikely they will be replaced by new coal power plants. The United States is moving rapidly to phase out coal for electricity production.

The energy industry in the United States is booming thanks in part to the use of fracking to produce more oil and natural gas. This industry employs over 2 million people, many in high paying jobs.

By default, in the United States actions to address global warming have been informally assumed by the states. California, the largest state by population and GDP, is playing a leadership role in addressing global warming. On the basis of GDP, California would rank as the world's fifth largest economy after Germany if it were a separate country. Other states are also taking action. Texas, the U.S. leader in fossil fuel production and use, is also the U.S. leader in wind power with 22,600 MW of installed capacity. In 2018, an estimated 18% of Texas electricity was generated by renewable wind and solar power.

We should all be hopeful that California's and other state's green energy initiatives are successful. In September 2018, California's Governor Brown signed SB100 (Senate Bill 100) stating that California would get all its electricity from renewable sources by 2045. This is a bold initiative and one that is achievable, but at a price.

As California succeeds and overcomes obstacles in its effort to get 100% of its electricity from renewable sources, it would be an example that other states and other countries can learn from and emulate.

This new initiative is consistent with California's long history of leading the United States in dealing with environmental issues. In the 1960s, California addressed air pollution problems by establishing the State Air Resources Board and led the nation in setting emissions standards for motor vehicles. This has been followed by legislation and other efforts to promote the use of low-emission vehicles and to deal with global warming.

In addressing global warming, California can demonstrate leadership, but cannot solve this global problem unless other states and other countries take similar actions. California, by itself, contributes less than 1.0% of the world's greenhouse gas emissions.

California has the resources to achieve its green energy objective to produce all electricity used in the state from solar or wind sources. With the right incentives, the private investment required should be available. Experience has shown that free market competition is often better than subsidies and mandates. This has been forcefully demonstrated in the case of coal. When natural gas became a cheaper fuel, no government agency had to tell utilities to burn less coal.

California has the land area, sunshine, and wind needed to deploy renewable energy. In addition, thanks to the Sierra Nevada Mountain range, the state has or could create the reservoirs needed for pumped storage if that was the best solution to store large amounts of energy to match the variable amount of electricity available from solar and wind with fluctuating electricity demand. Perhaps battery or other storage alternatives will become economical in time to contribute to solving this problem.

California has three problems:

- It has the highest electricity rates in the United States except for Hawaii and Alaska. Rates are set by a regulatory agency, rather than by competition as in Texas. In Texas, competition forces suppliers to pass on the declining cost of electricity from natural gas and renewable sources. Reducing the cost to consumers in California would speed up the conversion to electricity from renewable sources.
- The state continues to waste money on a bullet train that is way over its initial cost estimate and considerably behind schedule. Funding needed to complete this project has not been secured. The bullet train was initially promoted as a green initiative because, even during construction the steel, cement, and fuel needed would be offset by planting millions of trees. Where and when will they plant all these trees? This project may never be completed. If it is, its operation will not be green unless the electricity used is from renewable sources. Recognizing the situation, in 2019 newly elected Governor Gavin Newsom decided to scale back the proposed San Diego-San Francisco

route and instead complete a demonstration project that would connect Bakersfield, Fresno, and Merced in California's central valley.[4]

- The state has a Cap and Trade system to tax CO_2 emissions. It is a complicated system that is being gamed by the oil industry—to the extent that the oil industry now promotes cap and trade. One problem is that the carbon fee—currently $US17/ton of carbon—is too low. According to the World Bank, the fee needs to be $US40–$US80 by 2020 and then higher to reach the Paris Agreement goals. Currently four-fifths of cap and trade offsets involve forest protection projects, but the claimed emission reductions have not materialized.[5] A simple fee on carbon applied at the mine or wellhead is easier to administer and harder to avoid.

These problems can be solved. California and other states will make mistakes along the way to doing something about global warming. We should learn from these mistakes and support California's efforts and those of other states in the absence of a federal effort.

China—Will it be the leader?

In spite of being the world's largest CO_2 emitter, China is also the world's leader in renewable energy and electric vehicles.

China is the world's largest source of greenhouse gases, accounting for about 30% of the global total. China's greenhouse gas emissions are about double the United States. This is a result of China's population, the largest in the world at 1.4 billion, compared to 326 million in the United States. On a per capita basis, China's emissions are about 38% those of the United States.

China has to meet the rapidly growing demand for energy for its very large population. The economy had an average growth rate of nearly 10%/year since 1978. China now has a modern industrial economy and rapidly rising living standards. Economic growth and rising living standards require more electricity, gasoline, and other forms of energy.

---

[4] Phil Willon and Taryn Luna, "Gov. Newsom Pledges to Scale Back High-Speed Rail and Twin Tunnels Projects in State of the State Speech," *Los Angeles Times*, February 12, 2019.

[5] Jacques Leslie, "Cap and trade isn't cutting it," *Los Angeles Times*, January 2, 2020. (Article based on a report by *ProPublica*, November, 2019.)

Coal is the largest source of energy in China. China consumes almost as much coal annually as all other countries combined. China has the world's third largest coal reserves after the United States and Russia and is now the world's largest producer and user of coal, mainly to produce electricity. Unlike the United States, China does not have access to a large supply of low-cost natural gas to use instead of coal.

China is building hundreds of new coal-fired power plants capable of generating a total of 259 gigawatts (GW) of additional electricity. This equals the entire existing U.S. capacity of coal-fired power plants of around 266 GW.

In 2016, China imported more oil than the United States to become the world's largest oil importer. China is the fifth largest importer of natural gas, mostly in the form of Liquified natural gas (LNG).

China signed the Paris Agreement and has a national program to develop renewable energy sources. China's national strategy is to combat both global warming and air pollution. Another goal is energy security, to reduce its dependence upon energy imports. Pollution is largely due to the use of coal to produce electricity for industrial production and home heating. Vehicle emissions are a rapidly growing air pollution problem in urban areas. Severe air pollution also leads to contamination of food and water supplies, a major concern in China. China's cities are among the world's most polluted.

By many measures, China is also the world's leader in combating climate change despite being a large consumer of coal and oil. China has major efforts underway to develop renewable sources of energy, and carbon-free nuclear, and to replace gasoline and diesel vehicles with electric vehicles.

China has the world's largest installed capacity of hydro, solar, and wind power. In 2017, renewable energy comprised 37% of China's total installed electric power capacity, the vast majority from hydroelectric sources. China's investment in renewables grew by 100 times compared to 2005 so that China is by far the biggest investor in renewables with about one-third of global investments in renewable energy today.

China's automobile industry is rapidly turning all-electric. They produce more electric vehicles than the rest of the world combined. China not only has the biggest global car market based upon units sold, but is also the world leader in electric cars, with sales exceeding 1 million electric vehicles in 2018. China also has millions of electric bicycles and scooters. China is replacing millions of taxis, buses, and trucks with electric models. It is the world leader by far in producing electric buses and trucks. Its ambition is to make this a major export market.

China is the largest solar market in the world and has an installed capacity of around 130 GW, far greater than the United States at around 60 GW, and Japan at roughly 46 GW. The largest solar plant in the world at the moment is China's Tengger Desert plant with a capacity exceeding 1500 MW. China maintained its position as a wind energy powerhouse, installing 19.7 GW in 2017, while the European Union added 15.6 GW of capacity. The United States installed a little over 7 GW of capacity.

China is now leading the world in construction of new nuclear power plants. As of September 2018, China has 46 nuclear power plants in operation with a capacity of 42.8 GW and 11 under construction with a capacity of 10.8 GW. Beijing sees nuclear energy as an important baseload power source that is available, economic, and reliable. Nuclear power plants can become a major export business for China as well.

We can see that China is trying hard to expand renewable energy sources and reduce its dependency on coal and oil imports. It remains to be seen if China can grow renewable energy sources and nuclear power fast enough to reduce its coal use.

India—large population, little energy

India is the third largest energy user in the world, after China and the United States. Like China, India has a huge population, 1.3 billion, which is expected to exceed China's population in a few years.

India's problems are similar to China's. India depends on coal, its main domestic energy source, to meet its increasing demand for electricity. It has energy security issues, having to import most of its oil and natural gas. In addition, India has a very severe pollution problem, primarily from burning coal and from vehicle exhausts. According to the World Health Organization, 9 of the 10 most polluted cities in the world are in India.

Because of India's lower standard of living, on a per capita basis India's energy use is much lower than in China and the United States. India's per capita energy use is increasing rapidly from this low base as the standard of living in India improves.

After China and Germany, India is the world's third largest coal consumer. India has the world's fourth largest coal reserves, India's most abundant energy resource.

About 80% of India's oil is imported. India has a shortage of domestic natural gas and is increasing LNG imports. It is questionable if India can expand its renewable energy capacity fast enough to reduce its fossil fuel use.

India is going through an interesting transition. It is very dependent upon coal-fired power plants for its electricity production. Its problem is that new renewable energy is now cheaper to build than continuing to run most of India's existing coal-fired power plants.[6] The cost of renewables has decreased about 50% in the past 2 years. About 65% of India's electricity from coal power is being sold at higher rates than electricity from new renewable power plants. Much of the electricity from coal plants is sold under long-term purchase-power agreements. In addition, almost all of India's coal power plants violate the country's new air pollution standards. Retrofitting these plants to meet the new standards will make electricity from coal plants even more expensive compared to renewables. Stranded assets—plants that are no longer economical to operate—are already a problem for India's coal plants.

In addition, uncertainties over the future cost of fossil fuels are causing investors to be reluctant to invest in new coal plants. Renewables are increasingly a better investment opportunity for electricity production in that they do not have a fuel cost that would be uncertain. This is a trend across Asia and may be deterring some investments in coal plants.

Japan—strong technological capabilities

In 2017, Japan was one of only four major countries that reduced its greenhouse gas emissions. Because of Japan's unique position, its leaders are most concerned about energy security and want to reduce its dependence on imported fuels. Japan is an island chain with few natural resources, and is dependent on imports for almost 90% of its fossil fuels. Energy imports are a major contributor to Japan's trade deficit. Seventy-three percent of Japan is mountainous with most of its population concentrated in about 27% of its land area. Japan is very urbanized with high population densities. Limited land area inhibits the development of solar and wind farms.

[6] Silvio Marcacci, "India's Coal Power Is About to Crash," *Energy Innovation*, January 30, 2018.

Since the shutdown of Japan's nuclear reactors following the Great Eastern Japan Earthquake that caused the Fukushima nuclear accident in 2011, Japan had to increase its use of coal-fired power plants to meet the need for electricity. Prior to the accident, 30% of Japan's electricity came from 50 nuclear power plants. To offset the nuclear shutdown, Japan had to use more imported coal that increased the country's CO_2 emissions. The shutdown also led to higher electricity prices.

Japan is in the process of restarting its nuclear power plants after completing safety reviews. Japan is still counting on nuclear power to provide about 20% of its electricity long-term.

Japan's long-term goal to cut emissions 80% by 2050 was approved by the government. Japan is putting a lot of emphasis on conservation and efficiency improvements as well as investing in solar and wind energy.

Russia—may not be a player

Russia, the world's fourth largest emitter, is responsible for about 5% of the world's greenhouse gas emissions. Russia has little incentive to pursue renewable energy and reduce fossil fuel production, due to its abundance of fossil fuels and its dependence upon oil and gas exports. Russia was one of the 197 countries that signed the Paris Agreement in 2016 and ratified the agreement in 2019. This places Moscow in another highly visible leadership position, filling a void created by a U.S. withdrawal.[7]

Russia is one of the world's largest producers and exporters of fossil fuels. Russia's government is very dependent upon energy exports. It is estimated that about half of government revenues are from oil and gas exports.

Russia has the world's largest natural gas reserves and the second largest coal reserves after the United States. It also has large reserves of oil. The Russian economy is very dependent upon fossil fuel production and exports. Russia is also a major producer and exporter of steel and aluminum, both energy-intensive products to manufacture.

---

[7] Sabra Ayres, "Russia's Climate Opportunity: Moscow is Poised to Fill Void after U.S. Departure from Paris Agreement," *Los Angeles Times*, February 16, 2019, p. A2.

Russia is one of the world's largest producers of electricity from nuclear power. Unlike the United States, Russia continues to improve its nuclear technology and to build new nuclear power plants. A goal is to reprocess nuclear fuels to minimize waste and to recycle uranium and plutonium from spent reactor fuel into new fuel. Russia is also developing nuclear power as a major export industry. Russia has nuclear power plants under construction in several countries.

Nationally, large state-owned fossil fuel companies support Russia's energy needs and wield considerable political power. It is unlikely that they would support an increased use of renewable energy. Russia will lag other major countries in reducing greenhouse gas emissions. In addition, Russia's northern location and weather limit its solar energy potential.

Efforts by other countries to reduce fossil fuel use should eventually affect Russia's fossil fuel export markets and reduce Russia's fossil fuel exports and income.

Observations

There are no easy solutions. Every country has unique problems they must solve to reduce greenhouse gas emissions.

There are a large number of programs underway throughout the world to reduce fossil fuel use through greater use of solar, wind, and nuclear power, and more efficient use of energy. These efforts need to be expanded and accelerated if the world is to limit global warming to $2.0°C$ or less.

CHAPTER 12

Why is global warming such a difficult problem to solve?

Why is it so difficult to actually do something about global warming? Why cannot we limit global warming to 1.5°C as the Intergovernmental Panel on Climate Change (IPCC) recommends? Even if all countries met their pledges under the Paris Agreement, their nationally determined contributions would result in 2030 estimated emissions of 40 $GmtCO_{2eq}$, compared to the IPCC estimate that emissions cannot exceed 20 $GmtCO_{2eq}$ to limit global warming to 1.5°C.

In the previous chapter we reviewed some of the challenges facing six countries that are major contributors to global warming. While each country's situation differs, there are a number of issues that are common to all.

The need for unprecedented, perhaps unachievable, global cooperation

Perhaps the biggest obstacle to dealing with global warming is the need for international cooperation, especially between the United States and China, the two largest emitters. A coordinated international effort is required with firm commitments for reductions. If only a few countries take action, others get a free ride and not enough will be done to solve the problem. There has never been an example of this level of cooperation needed, or for the long-term coordinated effort that will be required.

This is a well-known economic dilemma first referred to as the *Tragedy of the Commons* by economist Garrett Hardin in 1968.[1] He cited an example of a shared resource (in our case, the atmosphere, in his case, the village commons, or shared pastureland). The tragedy occurs when each individual seeks to maximize his or her use of the "commons," with the inevitable result that overgrazing destroyed the common resource. It is

---

[1] Garrett Hardin, "The Tragedy of the Commons," *Science* 162, no. 3859 (1968): 1243−1248. doi:10.1126/science.162.3859.1243.

likely that some countries will seek to ease their commitments to minimize or slow negative effects on their economies. They will seek longer timeframes for action.

As mentioned elsewhere, we should acknowledge the global inequities in energy use. Poor regions such as Africa and India need to be allowed to increase their per capita energy use to provide more electricity for home and industrial use. They will need to substitute natural gas for firewood and provide gasoline for more motor vehicles or acquire electric vehicles.

The level of domestic and international cooperation to effectively address this problem and fairly share the associated costs and sacrifices may be beyond the current capabilities of today's domestic and international government institutions. Breakthroughs in international understanding and cooperation probably will be required.

Fossil fuels are heavily subsidized

As discussed in more detail in Chapter 15, Can It Be Done?, fossil fuel use is heavily subsidized. An estimated $US5.2 trillion/year is spent on subsidizing fossil fuel use. This includes tax breaks as well as cash subsidies in some countries for gasoline and electricity. The biggest subsidy by far is the ability to discharge greenhouse gases and other pollutants into the atmosphere essentially for free. The cost and damage related to these emissions, global warming and air pollution, is paid for by the general public, not by the generators of these emissions.

These subsidies promote the use of fossil fuels and delay the transition to renewable sources of energy by making fossil fuels less expensive than they should be when compared to renewables. In Chapter 14, Action Plan: Efficiency, Power, Transportation, and Land Use, we discuss the need for a carbon fee to offset fossil fuel subsidies. Remove the subsidies and fossil fuels would be more expensive and the transition to renewable wind and solar energy sources would be accelerated.

Educating the public

Educating the public and getting widespread acceptance of what is required to actually do something about global warming is challenging

Why is global warming such a difficult problem to solve? 165

but essential. There are several factors that work against getting public support. People have a tendency to believe what they want to be true. They do not want to be reminded of a growing global threat. There is a general belief that catastrophic events are rare and unlikely to affect them. It is human nature for people to want to "tune out" big problems.

There is an absence of comprehensive, readable, accessible information about global warming and what should be done about it. Much of the information available is complicated and written for scientists. People tend to rely on limited sources of information: the people they trust, newspapers, TV, or the Internet. The issue is further confused by the amount of disinformation being disseminated by those associated with the fossil fuel industry to confuse the public and promote the idea that global warming is not an urgent problem. Also, it is human nature that people give overwhelming priority to short-term over longer-term issues. The importance of a future problem is heavily discounted compared to short-term problems.

In the case of the United States, the government has not acknowledged that global warming is a serious problem that needs to be addressed immediately.

The media have not dealt fairly with global warming

In 1988, NASA scientist James Hansen testified before the U.S. Senate that global warming was potentially a serious problem and was caused by the accelerating use of fossil fuels worldwide. This was front-page news for the *New York Times*. At the time, President George H.W. Bush made a campaign pledge to combat the greenhouse gas effect.[2] Nothing happened. Thirty years have passed, 30 years lost that could have been put to good use to reduce greenhouse gas emissions and take other proactive steps. Why?

A good share of the blame falls on the media, particularly print and television and especially in the United States. Why? One reason was given by MSNBC commentator Chris Hayes, who wrote, "Every single time we have covered climate change, it has been a palpable ratings killer."[3]

[2] Mark Hertsgaard and Kyle Pope, "The Media are Complacent While the World Burns," *The Nation*, May 6, 2019, p. 15.
[3] Ibid., p. 12.

Apparently the demands of money and program ratings outweigh the media's responsibility to inform the public of impending crises, as important and unpopular as that role might be. Internationally, the media have been more proactive, so it is disappointing that media in the world's second highest emitter of greenhouse gas have been less responsive.

A partial explanation is due to the fact that newspaper and television are facing stiff competition for advertising revenue from the Internet. This has led to staff reductions, including investigative reporters. What is less well known is that the big fossil fuel companies have taken advantage of the situation. For decades the oil companies have known of the potential for global warming caused by their products, but have denied it. Even as long ago as 1970, ExxonMobil paid *The New York Times* to publish pieces called "op-ads" (as opposed to op-eds), to present views favoring production and sale of its product. Many highly respected news organizations—such as *The New York Times* and *The Washington Post*—have received hefty payments from ExxonMobil, Shell, Chevron, and the American Petroleum Institute to publish "advertorials" that they have written.[4]

The oil companies have also created a web of misinformation that casts doubt on legitimate scientific data. These efforts attempt to brand scientific results as "theory" rather than fact, or to minimize the consequences by suggesting that "negative impacts are so far in the future that surely solutions will be found—no need to worry."

In 2006, journalist Mark Hertsgaard discovered a link between the tobacco industry's lies about the dangers of cigarette smoking and the fossil fuel industry's misinformation about global warming. Physicist Frederick Seitz had received millions of dollars from R.J. Reynolds Company to falsify smoking risks. When the U.S. Surgeon General's report on the dangers of smoking dried up this lucrative business, Seitz became a climate expert overnight and collected large fees from oil companies to become a leading propagator of climate misinformation.[5]

It is time for the U.S. media to take global warming seriously. The media should cultivate staff members who are or who can become knowledgeable regarding climate science. Do not waste print or airtime on unqualified naysayers who want to present "alternative views." That is the thing about science; there is only one right answer.

---

[4] Amy Westervelt, "Sleeping with the (Greenwashing) Enemy: Why Are the *New York Times* and the *Washington Post* Producing Ads for Big Oil?" *The Nation*, May 6, 2019, p. 18.

[5] Hertsgaard and Pope, *The Nation*, p. 16.

Why is global warming such a difficult problem to solve? 167

News should be directed at critical regions. Coastal areas are facing sea level rise. In the heartland, cities are being destroyed by tornadoes, unusual flooding, and forest fires. Put the blame where it lies: lax government regulations, utilities companies fighting change, and oil company greed.

As one writer put it, "Those CEOs and political leaders who deny the well-established science of climate change should be tried for crimes against humanity."[6] This remark is a little extreme, but reflects the growing public concern that time for action is slipping away with no significant reduction in emissions occurring,

Public uncertainty

Some people do not believe that global warming is even real. They see no evidence in their own lives that global warming is occurring and is a problem. People could see smog in Los Angeles and experience immediate health effects. Someone in Ohio does not get too excited about the sea level rising. There is enough confusion and misinformation to cause many to feel hopeless or to ignore the problem. Others believe in global warming, but think the problem is too difficult to solve, or that nothing can be done in their lifetime.

In discussing global warming, the public needs to know the facts, consequences, and alternatives available. There are positive signs that awareness of the problem is growing and actions are being proposed.

In a recent Pew Research report, 59% of U.S. adults say that climate change is currently affecting their local communities. 31% say that climate change is affecting them personally. Those living near the coasts are more likely to say that climate change is affecting their community. In spite of this public awareness, the 2019 Pew Research Center's survey of public opinion of the important issues facing the country lists climate change as the 17th of 18 major issues on a list of concerns of the American public. Most people worry about something else (see Fig. 12.1).[7]

Even if someone is aware of the problem and wants to do something about it, often they do not know where to start or what to do. There are

[6] Ibid., p. 21.
[7] Pew Research Center, "Public's 2019 Priorities: Economy, Health Care, Education and Security All Near Top of List," January 2019, https://www.people-press.org/wp-content/uploads/sites/4/2019/01/PP_2019.01.24_political-priorities_FINAL.pdf. Used with permission from Julia O'Hanlon, Pew Research Center.

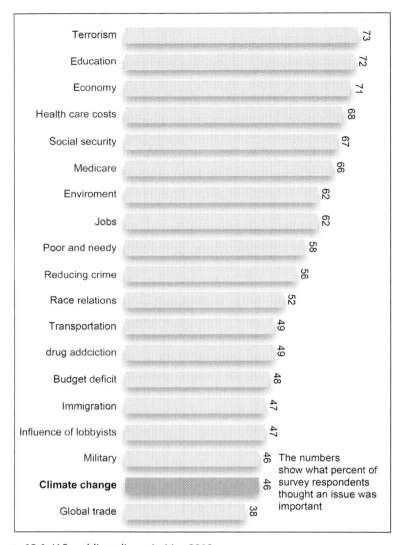

Figure 12.1 U.S. public policy priorities 2019.

personal actions that can be taken, such as using more energy-efficient light bulbs, or driving fuel-efficient, hybrid, or electric vehicles, using public transportation, or just driving less. But the most important step for an individual to take is to become informed and demand action from political representatives and business leaders.

Actions today such as curtailing the continuing use of fossil fuels may take 25, 50, or more years to have a visible effect. People may have to make significant sacrifices today to solve a problem that may not affect them very much. The problem is intergenerational. Steps need to be taken now to solve somebody else's problem in the future. This requires a long-term viewpoint that is too much to expect from populations that are struggling just to make ends meet, or those who are so focused on their day-to-day problems that they fail to pay attention to long-term health needs, do not save enough for retirement, etc. These people are being asked to sacrifice today to improve the lives of future generations.

Likewise, the impact of global warming will be very uneven. Some people, such as those living in desert areas or coastal plains, will be severely affected while many living elsewhere will be unaffected for a long time. Those who are not affected will be less concerned.

A positive message is needed

Asking people to make draconian sacrifices to fight global warming is a losing argument. Suggestions that people need to have fewer children, lower their standard of living, or change their traditional diet are unlikely to be accepted.

A better approach is to promote a more positive view of the future that will involve acceptable changes and some sacrifices to get there. Present a future that is worth the transition. We will reduce our energy consumption, but through steady improvements in conservation and efficiency improvements, not through painful austerity. We will have abundant energy to support acceptable and improving lifestyles, but it will not be from fossil fuels. Energy will be largely in the form of electricity from renewable sources. In addition, most air pollution will be eliminated, a big bonus. Some things will cost more, others less, but overall people will enjoy better health and a much cleaner environment.

Public support for government action

In democracies, governments cannot force the majority of the voting public to do something they do not support. Politicians will say they want to do something about global warming, but will not take actions that are not supported by the voting public, such as increasing gasoline prices. Politicians run for election every 2, 4, or 6 years, which limits their interest in solving problems that may be years in the future.

There are not any easy or painless solutions. To actually solve this problem will require some lifestyle changes, some sacrifices, and higher costs for fossil fuels for almost everyone living in energy-intensive societies. Until attitudes change, the average person will vote against any real pain such as higher gasoline prices. Effective actions will be politically unpopular and any politician who supports highly effective actions will risk losing an election. Voters prefer politicians who say they can fix a problem without any hard choices or sacrifices by the voting population, or who can convince the public that someone else will pay for it.

Advocates for addressing the global warming problem generally fail to grasp the full complexity of the issue when telling people what is really necessary to do something meaningful about it. They may recommend personal actions as mentioned above that are commendable (replace light bulbs with LEDs, drive a fuel-efficient vehicle, etc.). While these are useful first steps, much bolder actions are required to make a difference.

Why it is hard to replace fossil fuels?

To reduce greenhouse gas emissions to "net zero" it is necessary to eliminate the use of fossil fuels as well as make changes in agriculture and land use. As stated earlier, agriculture and land-use changes are responsible for about 25% of greenhouse gas emissions. For any fossil fuel uses that cannot be eliminated, their greenhouse gas emissions have to be offset by natural or artificial means to remove CO_2 from the atmosphere. This includes reforestation to increase the earth's natural carbon sink, or removing CO_2 from the atmosphere by artificial means using carbon capture and storage. Captured CO_2 would be stored underground, for example, in depleted oil wells. Most of the discussion focuses on CO_2 emissions from fossil fuel use, with CO_2 making up about 76% of greenhouse gases (see Table 2 in Chapter 6: How Do We Know Man-Made CO_2 Is the Issue?). Other greenhouse gases, methane, nitrous oxides, and fluorocarbons, have to be reduced as well. Agriculture is the source of most methane and nitrous oxide emissions. Fluorocarbons are from industrial processes. Reducing these other greenhouse gases will not be easy.

Fossil fuels are a huge global industry. There will be considerable resistance from the fossil fuel industry to the transition to renewable energy sources. The exploration, recovery, transportation, refining, and use of fossil fuels are a major global business. This industry will not volunteer to go out of business. It already has decades of denying that fossil fuels cause climate change and of funding individuals and groups that attempt to discredit scientific findings.

Modern economies depend on the use of abundant, low-cost, fossil fuels. Abrupt changes that risk lowering the growth of the economy or that increase unemployment are unlikely to be accepted by the public. We need to find a fairly smooth transition from today's fossil fuel-dependent economies to a future more reliant on renewable sources of energy.

Effective action to slow down or stop global warming could lead to some unemployment and perhaps lower gross domestic product growth. Some countries dependent upon fossil fuel exports would see their government and private sector revenues decline. There may not be an easy transition from current energy sources to alternate energy sources.

Even if the overall cost of addressing global warming is less than the benefits, the costs and benefits are unevenly distributed. Those who pay the price are unlikely to be the same people who receive any benefit. There will be winners and losers to deal with.

Solving technical challenges

We have to start solving the urgent problem of global warming with the technology available today. Future technology may help, but we cannot wait for the possibility that some better solution will be available sometime in the future. Fortunately, what is available today is adequate. There are some difficult problems to be solved. However, these problems are being addressed and solutions should be found for most, if not all, of the problems listed below.

Today's technology alone may not provide the best solutions to global warming. However, we are headed in the right direction with future energy efficiency advances, more efficient solar and wind power generation, improved energy storage technology, and other means of reducing dependence on fossil fuels.

Economies of scale and new technology are driving forces reducing the cost of electricity from wind and solar. Already, electricity from these sources is cheaper than fossil fuel alternatives in many locations. It is only a matter of time before solar and wind will be the low-cost source of electricity for most locations.

The biggest technical challenges are:

- *The intermittent nature of renewable electricity from solar and wind*

    This problem was discussed in Chapter 10, Energy Alternatives. Possibly some nontechnical solutions will help, such as time-of-day pricing for electricity from solar sources. If electricity at midday is much less expensive than at other times, would consumers time some of their electricity use to take advantage of lower cost electricity? We will need further cost and technical improvements with energy storage to deal with the intermittent nature of solar and wind energy. Costs and the technology are improving and a number of advanced energy storage options are being developed and installed.

- *The need to find a cost-effective, carbon-free source of liquid fuels to replace gasoline, diesel, and aviation fuels*

    Biofuels have serious disadvantages. The move toward electric vehicles will help. However, it is unlikely that electric vehicles will replace all vehicles using oil for an energy source. Transportation remains a major challenge. Synthetic fuels produced from captured CO_2 using renewable energy are technically feasible and may become cost-effective alternatives.

- *The need to rethink electric transmission and distribution systems*

    The past model was a distribution network designed for one-way flow of electricity from large central power stations to widely distributed customers—households, commercial, and industrial customers. The new grid needs to be more intelligent and more like the Internet. Renewable energy producers will be widely distributed, in addition to large utility-scale producers. Depending upon time of day, electricity can flow in either direction from rooftop solar installations, for example, to local and distant customers. Refer to the discussion in Chapter 10, Energy Alternatives, concerning smart grids.

- *Carbon capture*

  There are several alternatives under development to capture CO_2 as it is emitted from a coal-fired power plant and store it underground in locations such as depleted oil fields. Other schemes would capture CO_2 in the atmosphere and convert it into useable liquid fuels or store it forever. Refer to Chapter 10, Energy Alternatives, for a discussion of this technical problem.

- *Nuclear: still an option?*

  Nuclear power has been a reliable source of carbon-free electricity. Unfortunately, nuclear power plants under construction in the United States are very expensive and have chronic problems with construction delays and cost overruns. Some promising new alternatives for the design of nuclear power plants are being developed. However, it will take a long time to demonstrate and deploy any new nuclear plant designs in sufficient numbers to make a difference. The present status and future for nuclear power are discussed in more detail in Chapter 13, Some Success and Failures.

The need for strong economies

Strong and growing economies are necessary to fund private sector and government programs to transition to renewable sources of energy. If the economy stumbles, the ability to make important investments in renewable energy and other worthwhile programs will be compromised. The private sector economy is the source of government revenues. The private sector economy is also the primary source of new jobs. Government jobs require fund transfers from the private sector to pay for them. Higher fossil fuel prices will have some impact on the growth rate of the economy. This is unavoidable. Alternatively, major investments in smart grids and other projects could stimulate growth and employment. These changes will take place over 30 years or more and need not have a large, adverse impact on the economy.

Required government investments will need additional revenue. The major potential source of additional revenue should be a carbon fee. Without carbon fee revenues, infrastructure projects and other spending associated with the switch to renewables would be financed by increased government deficits and debt.

In the case of the United States, prior to 2019 it has been able to increase its deficits and national debt without any obvious negative effects. Lower interest rates have offset the growth in the U.S. national debt so that interest expense has actually fallen. The U.S. federal deficit was over $US1 trillion in the current fiscal year, before the massive stimulus to offset the Corona virus impact on the economy. There have to be limits to deficit spending and increasing debt, and at some point unchecked entitlement spending could become a problem.

Understanding climate change skepticism

There is probably very little that one can do to otherwise convince those whose minds are made up if they are not receptive to a discussion. No scientist disputes the fact that the CO_2 concentration in the atmosphere is increasing. This is an accurately measured trend and is not in dispute. The source of increasing CO_2 concentration can be measured and is known to be due to human activity. The earth's temperature is rising as well. There is a close relationship between CO_2 concentration in the atmosphere and temperature increase. The close relationship between CO_2 concentration and the earth's temperature has been true for over 800,000 years. The greenhouse effect under which CO_2 and other greenhouse gases act to increase the earth's temperature is also a fact, not theory. It has been verified by scientific analysis and experiments for a long time.

The main argument is whether the temperature increase is caused by man-made CO_2 emissions, natural variations, or some combination of both. The same is true for rising sea levels. Rising sea levels are caused by increasing ocean temperatures causing thermal expansion as well as water flowing into the oceans from melting glaciers and ice caps. Deniers will argue it is a natural variation. Some of the most common arguments follow.

Many climate deniers do not see evidence of global warming in their lives. It cannot be happening. We have had an exceptionally cold winter, how can there be global warming? I have been going to the beach for 20 years and do not see any evidence of rising sea levels, etc.

Some argue that increased CO_2 concentration and warmer atmosphere and oceans are actually beneficial. Increased CO_2 concentration promotes plant growth, leading to a greener earth and higher crop yields. Higher ocean temperatures increase humidity which they claim is beneficial to plant growth.

Some are paid advocates for an organization or industry. That does not mean that they cannot have valid arguments. But their sponsors and sources of funding should be considered when evaluating their arguments.

Other climate deniers argue that the changes required to stop global warming are too draconian or that there is nothing that we can do anyway. Why bother?

Still others argue that the money that would be spent on reducing global warming would be better spent on other purposes such as fighting disease or improving water quality.

Some believe that global warming is a hoax being promoted by a lot of scientists working for governments and universities. It is a way to further their careers, guarantee employment, or get grant money. They cannot be trusted.

Most of those who disagree or minimize the threat of global warming also tend to minimize the effects of higher temperatures, flooding, or severe weather.

In general, climate deniers ignore the elephant in the room. Can the world continue to discharge 55 GmtCO$_{2eq}$ (refer to Fig. 9.1, Chapter 9: What Would It Take to Reach Net Zero?) into the atmosphere indefinitely without unacceptable consequences? If not, what are the limits and when do we need to deal with this problem? Could anyone believe that there is not any limit? There are abundant fossil fuel reserves to be tapped, more than enough to increase the CO_2 concentration in the atmosphere to 1000 ppm over the next 100 years or so, depending upon whose estimate is used. Would not one think that eventually this becomes a problem?

It should be possible to analyze and discuss the main disagreements between the IPCC and others who support taking action, and those who disagree. The facts are available to both sides of the argument. Rather than allow this disagreement to continue, there should be some effort to objectively assess the facts, and make conclusions in a form that is accessible to the concerned citizen. Perhaps the best way is for climate change deniers to submit their case in scientific papers that would be subject to peer review, as do other scientists.

Some climate change deniers would want us to believe that somehow there is some nefarious force that causes thousands of scientists in many disciplines and from dozens of countries to come together to promote a false set of facts, but for what reason? It makes no sense. Consider the IPCC. Its reports are compiled by thousands of atmospheric scientists, climate modelers, oceanographers, ice specialists, economists, and public health experts, mostly drawn from universities and research institutes. They work on a volunteer basis. IPCC reports are extensively reviewed by independent experts. The IPCC draws support from 195 member countries.

A large number of people are closed-minded and have already decided for a variety of reasons that global warming is not a real problem. Others are aware of global warming and may be concerned about this problem. However, they are subject to misinformation and do not know what, if anything, they should do or believe. They hear a wide range of opinions and isolated facts that can be confusing.

Recognizing political leaders can make mistakes

Money raised through a carbon tax or fee can be misspent. Politicians often favor specific solutions such as ethanol rather than letting competition decide the best solution based upon price and other considerations.

California's Cap and Trade program is an example. While the idea sounds good in theory (make emitters buy credits from firms that are reducing emissions and use the resulting income to invest in green technologies), the implementation can fall short. In California's case, the "Bullet Train" project (described in Chapter 11: Unique Problems of Major Contributors to Global Warming) to construct a high-speed rail system between Los Angeles and San Francisco is a good example. The $70-plus billion dollars being spent would be far more beneficial if used to construct solar farms to meet the State's 2045 goal for 100% renewable energy. There has been no comprehensive evaluation of the carbon reduction that would result from the bullet train, including the carbon being generated by the mammoth construction project, the land-use impacts, the manufacture of the running equipment, and the energy to operate the train itself.

Acknowledging that failure is a possibility

The earth's temperature is already about $1.0°C$ above the preindustrial baseline of 1850–1900. CO_2 levels have already increased from 280 to 410 ppm. Even if we start today with aggressive actions to reduce the use of fossil fuels, it is not unrealistic to anticipate CO_2 levels of about 500 ppm before we can reduce man-made emissions to a level that is equal to or below the level that can be absorbed naturally by the earth's atmosphere. Due to latency, the full effect of CO_2 in the atmosphere and oceans is not yet known. This would result in a further rise in the average global temperature, possibly as much as 2.0°C or more. This temperature change is enough to lead to major changes in the earth's climate such as further polar ice melting, sea level rise, increased storm activity, and droughts.

As mentioned previously, *failure is a possibility*. The degree of cooperation, foresight, and sacrifice needed may be beyond our current capabilities on a national or global basis. We may have to experience a severe crisis before effective actions are taken. We could see a substantial increase in the earth's temperature with serious effects on the earth's climate.

CHAPTER 13

Some successes and failures

Over the years, there have been some successes and failures in environmental and energy matters that offer lessons worth remembering as we proceed to address global warming. Successes will be discussed first.

The Permian Basin, a renewable energy powerhouse

The Permian Basin is a large oil formation in west Texas and New Mexico. It is about 300 miles long and 250 miles wide, covering about 86,000 square miles. This is the size of Minnesota, the twelfth largest state and almost half the size of California.

The basin has been producing oil for about 100 years. Thanks to fracking, a new drilling technique, oil and gas production in the basin has been increasing steadily. In May 2019, the basin produced 4.5 million barrels of oil and 15.0 billion cubic feet of natural gas. This is about 35% of U.S. oil production and 17% of natural gas production. The basin is now the world's largest oil field as measured by daily production, with oil output exceeding that of Ghawar, in Saudi Arabia.

As impressive as this is, the Permian Basin is becoming a powerhouse for renewable energy as well. Its potential to produce electricity from solar and wind farms could exceed the energy equivalent of its oil and gas production. Based upon the authors' estimates, the Permian Basin with its strong winds and sunny skies has the potential to produce more energy from wind or solar farms than is presently being produced by its oil and gas wells. The land area is large enough that the basin should be able to produce oil and gas, as well as renewable energy from both wind and solar farms, all at the same time. The basin's oil and gas reserves are finite and will eventually be depleted. However, its wind and solar energy potential will last forever.

Texas now has an installed capacity of 25,000 MW of wind power, most of it in west Texas. This is enough electricity to power over 6 million homes. In 2018, Texas produced almost a quarter of all wind energy

Reaching Net Zero.
DOI: https://doi.org/10.1016/B978-0-12-823366-5.00013-0
© 2020 Elsevier Inc.
All rights reserved.
179

in the United States. The state produced 69.8 million MWh (18.6% of Texas' total electricity from wind, more than from coal).

Texas has made some farsighted decisions that facilitate the development of renewable energy:

First, the Energy Reliability Council of Texas (ERCOT) completed a $7.0 billion infrastructure project to construct thousands of miles of transmission lines to connect the windy and sunny plains of west Texas to Dallas, San Antonio, and other load centers in the eastern part of the state. This connection gave west Texas access to major markets to justify the construction of large wind and solar farms.

Second, Texas is now the country's largest deregulated market for electricity. Texas deregulated its power system so that electricity providers have to compete for customers' business. This competition forces suppliers to pass on declining electricity costs to customers. So far, most of the price reduction is due to the displacement of coal by lower cost natural gas for electricity production. In January 2019, the average price of electricity in Texas was 11.68 cents per kilowatt-hour, less than the national average of 13.31 cents per kilowatt-hour. In California, another state committed to green energy, the average price was 19.90 cents per kilowatt-hour. Under Texas' deregulated market, the declining cost of wind and solar electricity will be passed on to customers in Texas and will speed up the use of renewable energy.

The state is also pioneering ways to integrate renewable electricity from wind and solar sources into the transmission and distribution system.

Plans were announced to build a 495-MW solar farm and electricity storage system to provide power to the oil fields in the Permian Basin. This will be the world's largest battery so far and the largest solar farm in Texas. Fracking and the expansion of oil and gas production require additional electric power to run pumps and other oil field equipment. Solar arrays with battery storage will provide the power needed to increase oil and gas production.

There are several new wind farms under construction, such as the 450-MW High Lonesome project representing a $600 million investment.[1]

Texas is rapidly becoming a model for renewable energy. It has demonstrated the value of thinking big by developing the transmission lines

---

[1] Mella McEwen, "Construction Begins on $600 Million West Texas Wind Farm, "*Midland Reporter-Telegram*, January 20, 2019, https://www.mrt.com/business/oil/article/Construction-begins-on-600-million-West-Texas-13542684.php.

Some successes and failures | 181

needed to carry renewable electricity from where it's best produced in west Texas to where it's needed.

If Texas can send renewable energy from west Texas to Dallas, why not to Chicago?

1970s oil price hikes

In the 1970s, the world was forced to respond to sudden and large increases in the price of oil. Although these price hikes caused recessions, the global economy recovered quickly and responded to price increases with greater conservation and improved efficiency measures. The oil price hikes of the 1970s are also a real example of the impact rising prices have on the demand for oil. There were actually two oil crises, one in 1973 and another in 1979, when the Yom Kippur War and the Iranian Revolution triggered supply interruptions and price increases in Middle Eastern oil exports.

Transportation is almost totally dependent upon oil-based liquid fuels. Oil prices had been very stable from the end of World War II until 1973. The price increased slowly from $2.77 per barrel in 1948 to $3.60 in 1972. Low and stable oil prices supported the increasing use of oil. From the end of World War II until the early 1970s, U.S. oil consumption increased about 4.7%/year from 5.6 to 16.8 million barrels/day. This growth rate came to an abrupt halt with the first oil embargo in 1973.

Low oil prices prior to 1973 resulted from the ability of the large global oil companies to negotiate favorable prices with the oil-exporting countries. These large oil companies were known as the Seven Sisters. They controlled global oil exploration, production, distribution, refining, and sales.

The Organization of the Petroleum Exporting Countries (OPEC) was founded in 1960 to give the oil-exporting countries more leverage in negotiating prices with the large oil companies. It was made up of 15 oil-exporting nations. It included all the oil-rich Arab countries in the Middle East and some Latin American countries. These oil-exporting countries had seen their oil incomes decline after past failures to negotiate better prices.

U.S. domestic oil production peaked in 1960 and the United States started to import more oil. This increased OPEC's leverage in setting oil prices. In October 1973, the Arab members of OPEC (OAPEC) proclaimed an oil embargo in response to the U.S. decision to resupply the Israeli military during the Yom Kippur war. OAPEC declared it would limit or stop oil shipments to the United States and other countries if they supported Israel.

In October 1973, OAPEC raised their oil price by 70% to $5.12 a barrel. Over the next 2 months, they cut production and stopped oil shipments. By the time the embargo was lifted in March 1974, oil was selling for about $12 a barrel, over three times its earlier price. The price of gasoline in the United States increased from $0.25 per gallon to over $1.00 per gallon. As a result, oil consumption in the United States dropped 20% rather quickly.

The 1973 embargo was a shock for most countries that enjoyed virtually unlimited supplies of cheap oil in the 1960s. The post-World War II economic growth came to an end with a global recession caused by abrupt oil price increases.

The second oil shock occurred in 1979 following the disruption resulting from the Iranian Revolution during which the Shah, a U.S. ally, was deposed by the religious leader Ayatollah Khomeini. Although global oil supply only decreased by about 4.0%, the price of oil increased to $39.50 per barrel. This price is more than $110 per barrel when adjusted for inflation using 2017 prices. The reduced oil supply and the price increase triggered recessions in the United States and other countries.

As a result of price increases, more oil reserves became economical to exploit and oil-producing countries expanded oil production and new oil fields were brought on line in Alaska and the North Sea. These new supplies weakened OPEC's negotiating position and eventually led to lower oil prices. Due to abundant global oil supplies, oil prices began a 20-year decline starting in 1980.

What were the consequences of these oil price increases? The end of the 1970s was the historic peak for global per capita oil consumption, which has been flat for 35 years at about 4.5 barrels per year per person (see Fig. 13.1).

Growth in global total carbon emissions went from about 5% annually before 1973 to less than 1% annually since then, due to energy efficiency and conservation improvements in industry and transportation (see Fig. 4.5). In 2019, the world emitted a record 37 Gmt CO_2 from fossil fuel consumption (see Fig. 6.4). If carbon emissions had continued to rise at their pre-1973 pace, they would have exceeded 100 $GmtCO_2$.

In reaction to the oil price hikes, the U.S. automobile industry responded with attempts to produce lighter, more fuel-efficient vehicles.

Some successes and failures 183

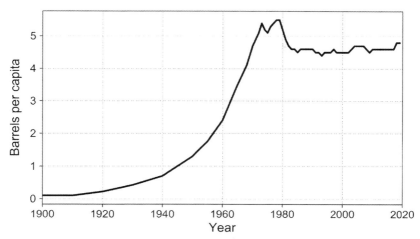

Figure 13.1 World oil consumption per capita.[2]

Japanese automobile companies increased their market share since they already had a number of smaller, more fuel-efficient vehicles. There were a number of innovations to improve fuel economy, such as the use of turbochargers, lighter weight materials, more efficient transmissions, front-wheel drive, and fuel injection. It is estimated fuel economy improved about 80%. These changes coincided with California's push for automobile emissions controls. The U.S. Congress passed emergency conservation measures including a nationwide 55 mph speed limit and a 1975 energy bill establishing mandatory fuel economy standards.

Real energy prices (adjusted for inflation), declined steadily for almost 20 years, reaching an all-time low of about $11.00 per barrel in 1998. Since then, oil prices peaked at about $120.00 per barrel in 2010 during the Arab Spring uprising, followed by a decline to about $50.00 per barrel.

Thanks to decreases in the real price of gasoline since about 1980, the U.S. public has favored larger, more powerful vehicles. In 2018, 70% of the vehicles sold in the United States were sports utility vehicles (SUVs) and pickup trucks. In the 1970s, following the first energy crisis, you could not give away a vehicle with a V8 engine.

[2] Source: World population by year from https://www.worldometers.info/ Oil consumption by year from Our World in Data, global fossil fuel consumption from 1800 to 2016. Data for 2017 to 2019 from EIA Short-Term Energy Outlook www.eia.gov/outlooks/steo/report/global_oil.php

Automobile emissions

In the 1950s and 1960s, smog became a major problem in cities such as Los Angeles. Smog was an obvious health hazard and smog's brown haze was clearly visible. The old joke was that Californians were uncomfortable breathing air they could not see. On smoggy days, many people had physical symptoms such as itchy eyes and sore throats. It was an easily seen health concern. Thanks to scientific research, there was a clear causal link between vehicle emissions and air quality. There were not any smog deniers.

California took the lead and established emission standards in 1966 for all cars first sold in California. The State of California established the California Air Resources Board in 1967 to deal with this and other air pollution problems. In 1970, the federal government passed the Clean Air Act and established the Environmental Protection Agency.

Early automobile emissions controls reduced exhaust emissions but also reduced engine performance and mileage. It took time for the automobile industry to perfect emission control systems that did not compromise performance. In the 1970s, the catalytic converter was introduced to control emissions while minimizing the effect on engine performance.

In addition, the oil industry had to eliminate the use of leaded gasoline because leaded gasoline degraded catalytic converters. The elimination of leaded gasoline was a major change and was completed in 1975. Besides eliminating a health hazard, unleaded gasoline reduced the periodic maintenance formerly required by internal combustion engines. Air quality regulations led to automobile industry innovations to reduce vehicle emissions and to the oil industry eliminating leaded gasoline. All this took time, effort, and major investments to modify automobile engines and to upgrade oil refineries. Standards were tightened over time, leading to further improvements.

California was the largest U.S. automobile market and had the clout to enforce emissions standards. No major automobile manufacturer could afford to concede the California market to its competitors. California's efforts soon resulted in national standards for the United States. Many other countries also implemented similar programs. Foreign automobile manufacturers had to meet U.S. standards for vehicles sold in the United States or forfeit a lucrative market.

Anyone who lived in California in the 1950s and 1960s knows the dramatic improvement in air quality that resulted and continues to this day. What makes this achievement even more remarkable is that in 1960 there were 7.8 million passenger cars, trucks, and busses in California, but by 2015 the number had grown to 34.3 million.

Hole in the ozone layer

In 1976, the U.S. National Academy of Sciences concluded that there was a link between the use of chlorofluorocarbons and depletion of the ozone layer. This finding brought about further research and measurements of the ozone layer. The primary cause of ozone depletion was the presence of man-made chlorine-containing gases, mainly chlorofluorocarbons used for aerosol sprays, packing materials, solvents, and refrigerants.

In May 1985, scientists with the British Antarctic Survey announced the discovery of a huge hole in the ozone layer over Antarctica. This led to immediate concerns that it could result in more harmful ultraviolet (UV) radiation reaching the earth.

Ozone (a molecule consisting of three oxygen atoms, or O_3) is a small constituent of the atmosphere but is responsible for most of the absorption of UV radiation. This reduces the amount of UV radiation that hits the earth's surface. An increase in UV radiation increases risks to humans for sunburn, skin cancer, and cataracts.

The ozone hole is actually not a hole but a thinning of the ozone layer allowing more UV radiation to reach the earth's surface. The largest ozone hole is seasonal and is in the Antarctic from September to December, the spring months in the southern hemisphere. There is a smaller seasonal hole over the Arctic. From a global warming perspective, increased UV radiation may increase the melting of sea ice in the Arctic and may have other harmful effects.

There was an international response to the ozone depletion problem. In 1985, 20 countries, the major chlorofluorocarbon producers, signed the Vienna Convention for the Protection of the Ozone Layer that determined how international regulations would be developed to control the use of ozone-depleting substances. In 1987, 43 nations signed the Montreal Protocol to freeze chlorofluorocarbon emissions at 1986 levels.

Industry cooperated and substitutes for chlorofluorocarbons were developed. At a meeting in London in 1990, it was agreed to phase out the use of chlorofluorocarbons except for a few small applications where suitable substitutes were not available.

In 1994, the United Nations General Assembly voted to designate September 16 as the International Day for the Preservation of the Ozone Layer, or "World Ozone Day," to commemorate the signing of the Montreal Protocol on that date in 1987.

Although there has been a well-coordinated international response to reduce man-made chlorofluorocarbons, the level of these substances in the atmosphere is not expected to return to pre-1980 levels for an estimated 40–50 years. This is due to the slow rate at which chlorofluorocarbons are naturally removed from the atmosphere. It is appropriate to remember this lesson, since CO_2 stays in the atmosphere for a long time as well.

Cigarette smoking and cancer

We mentioned cigarette smoking and lung cancer in Chapter 1, Introduction, as an example of latency. Tobacco was first introduced to Europe by Christopher Columbus. But it was not until 400 years later that cigarette smoking took off, after the development of machines to mass-produce cigarettes. Cigarettes became a cheap and easy way to smoke tobacco. In the 1960s, about 42% of the U.S. population were smokers, about 50% of men and 30% of women.

That cigarettes were a health hazard was fairly well known; cigarettes were referred to as "coffin nails," for example. However, the cigarette companies went to great lengths to obscure the linkage between cigarette smoking and lung cancer and other diseases. The connection between cigarette smoking and lung cancer was established scientifically starting in the 1950s. However, it took the U.S. Surgeon General's report, issued in 1964, to clearly establish that cigarette smoking caused lung cancer. This led to a number of actions to reduce the consumption of cigarettes. Later reports highlighted the dangers associated with secondhand smoke.

In 1964, labeling cigarette packs was required, stating "Caution: cigarette smoking may be hazardous to your health." Other measures followed. Smoking was banned on airplanes and there was a move to establish smoke-free zones in public areas.

In 1988 California voters approved an increase in cigarette taxes. Other states followed suit. Today, the average tax on a pack of cigarettes is $1.68. Actual taxes vary by state from about $0.20–$4.35 per pack. This greatly increased the cost of smoking cigarettes. In 1993, the U.S.

Food and Drug Administration assumed jurisdiction over tobacco products and declared nicotine a drug.

Several states sued the major tobacco companies mainly to compensate the states for their increased health-care costs associated with cigarette smoking. The *Tobacco Master Settlement Agreement* was entered in November 1998 between the four largest tobacco companies, and the attorneys general of 46 states to recover their tobacco-related health-care costs. The companies agreed to curtail or cease certain tobacco marketing practices, as well as to make, in perpetuity, various annual payments to the states to compensate them for some of the medical costs of caring for persons with smoking-related illnesses. The general theory of these lawsuits was that the cigarettes produced by the tobacco industry contributed to health problems among the population, which in turn resulted in significant costs to the states' public health systems. In exchange the companies would be freed from class-action suits and litigation costs would be capped.

In spite of these efforts, even today about 18% of the people in the United States smoke and cigarette smoking is still the leading cause of preventable disease and death in the United States.

There have been a number of policy failures that have implications for dealing with global warming. The following four are worth reviewing to see what we should learn from past failures.

Europe's push for diesel vehicles

Europe is the world's largest market for diesel-powered light vehicles and accounts for about 70% of all diesel vehicles sold worldwide. Because of the incentives associated with diesel vehicles in Europe, around half of the vehicles in the region are diesel-powered.

For a long time, Europe has favored diesel-powered vehicles over those with gasoline engines because it was believed that diesel vehicles got better mileage and thus emitted less CO_2 than the equivalent gasoline-powered vehicles. This helped Europe meet CO_2 reduction targets established in the Kyoto Protocol of 1997. European governments have spent billions subsidizing diesel fuel to make it cheaper than gasoline, plus adding other incentives such as taxing new diesel vehicles at lower rates than gasoline-powered vehicles.

Recently, it has been determined that the emission advantages of diesel vehicles are less than originally thought. In 2015, Volkswagen was caught cheating on the reported emissions test required by the United States. Volkswagen intentionally programmed diesel engines to activate their emission controls only during laboratory emissions testing in order to meet U.S. standards for nitrous oxide emissions. These events led the public to be more concerned about the health effects associated with diesel vehicles.

Unlike the United States, in Europe only CO_2 emissions were regulated. In addition, CO_2 emissions in Europe are based upon the average vehicle weight of a manufacturer's fleet of vehicles. The heavier a manufacturer's average vehicle, the higher the level of CO_2 emissions that were allowed. There is little or no incentive to reduce vehicle weight. These factors favored heavier diesel-powered vehicles. Lower prices for diesel fuel led to people driving more miles, offsetting some of the perceived efficiency advantages of diesel-powered vehicles.

Diesel-powered vehicles produce high levels of nitrogen dioxide and fine particulates that increase pollution in urban areas. In February 2018, Germany's Federal Administrative Court issued a landmark ruling that required older diesel vehicles to be banned from congested downtown streets in acutely polluted locales. The ruling upheld bans proposed in Stuttgart and Düsseldorf, two of Germany's smoggiest cities, and set a precedent that applies across the country. Other European cities are also considering bans of diesel-powered vehicles in city centers.

This has led to a drop in the purchase of diesel-powered vehicles and a sharp drop in their resale value.

Nuclear power in the United States

There are still 98 nuclear reactors in 60 nuclear power plants in 30 states operating in the United States, producing about 18% of the nation's electricity. The United States is presently the world's largest producer of electricity using nuclear power. However, almost all of this power is produced by plants built between 1967 and 1990. The average age of U.S. commercial reactors is about 37 years. The oldest operating nuclear power plant in the United States is Nine Mile Point in New York, which entered commercial service in 1969. Most plants already in operation are nearing the end of their service lives or are becoming uneconomical to operate in competition with low cost, but CO_2 emitting, natural gas. The amount of electricity generated by nuclear power plants is declining in the United States.

In the 1950s, the U.S. government under the Eisenhower administration initiated a program, Atoms for Peace, to harness nuclear energy for the production of electricity. This was a genuine "*Swords into Plowshares*" project with a lot of support from the federal government and the private sector. It was well-funded and well-staffed by thousands of experienced scientists and engineers who worked in nuclear weapons programs, in the utility industry, and in the Navy's nuclear propulsion program. The U.S. Navy nuclear power program was for submarines and aircraft carriers. The safety and reliability record for these nuclear-powered vessels is outstanding. The carrier *Enterprise* was in service for over 50 years.

In the beginning, it was believed that electricity produced by nuclear power plants would be inexpensive. There would be an unlimited supply of cheap electricity. What went wrong?

The United States has largely lost its capacity to design and build commercial nuclear power plants. The leading producer of commercial nuclear reactor systems in the United States, Westinghouse, sold its nuclear business to Toshiba in 2006. In 2017, Westinghouse's nuclear business, now owned by Toshiba, declared bankruptcy. China is now the leader in commercial nuclear power with 46 nuclear plants in operation and 11 under construction.

The public has a great fear of radiation and radioactive waste products. It has been impossible to convince the public that it is safe to live near a nuclear power plant. The utility companies built their plants in their service areas on land they owned with access to large sources of cooling water. As a result, many of these plants were built near population centers and were opposed by those living near the plants. The nuclear reactor failures at Three Mile Island (United States) in 1979, Chernobyl (USSR) in 1986, and Fukushima (Japan) in 2011 confirmed the public's worst fears of nuclear power.

Initially, utilities were guaranteed a fixed rate of return on their investment in nuclear power plants even if these plants had cost overruns. Any cost overrun from these very expensive nuclear power plants was added to a utility's rate base. They were able to petition for rate increases if needed to get the guaranteed rate of return on their investment. This arrangement ended with deregulation of the utility industry under which they were only paid for the electricity they produced at competitive rates. This transferred major financial risks from the utilities' customers to the utilities.

There is not any cost associated with greenhouse gas emissions. Coal- and gas-fired fossil fuel plants paid little or nothing for the emission of huge quantities of greenhouse gases and other pollutants into the atmosphere. Nuclear plants did not get any financial benefit from being emissions-free. To be fair, nuclear power plants did have some significant government subsidies.

Many utilities did not have the management or technical expertise to manage the construction and operation of the more complex and expensive nuclear power plants. This contributed to cost overruns, licensing delays, and operating problems at some plants. Numerous changes in the Nuclear Regulatory Commission safety requirements to address growing safety concerns made the licensing, construction, and operation of nuclear plants increasingly difficult and costly. New safety requirements that were applied after the start of construction led to expensive retrofits causing additional cost overruns and schedule delays.

The U.S. government pledged to receive and safely store radioactive spent fuel. Despite decades of work and billions of dollars spent, the U.S. government failed to come up with a politically acceptable solution for the disposal of spent nuclear fuel. It is presently stored on site in large water-filled pools or dry casks at each nuclear power plant. It is noteworthy that France, a country smaller than the state of Texas, has had no difficulty in safely processing and storing its radioactive waste.

France was also an industry pioneer and leader with 58 nuclear power plants in operation. This resulted from the French government deciding in 1974, just after the first oil shock, to expand the country's nuclear power capacity to become more energy-independent. France produces about 75% of its electricity from nuclear power. The French nuclear power plants are close to 40-years-old and some of those plants must either be retired or have expensive upgrades to extend their operating lives. There is one new nuclear power plant under construction in France. It is behind schedule and over budget. The billions it will cost to make such upgrades is money that could be spent developing the country's green energy program. The decline in nuclear power will result in France relying more on coal and natural gas.

Recent experience in the United States is not positive. Only one new nuclear plant with two reactors is under construction. No additional nuclear power plants have been ordered.

In 2008 two new nuclear plants started construction. These were the first new nuclear plants in the United States in three decades. The Vogtle plant in Georgia with two reactors is still under construction. This plant is behind schedule and over budget. It is about 60% complete with current estimated completion dates of 2021 and 2022 compared to the original dates of 2016 and 2017—an additional 5 years. The current cost estimate is $23.0 billion compared to the original estimate of $14.3 billion. So far, the utilities building this plant are committed to completing it.

The V.C. Summer nuclear plant with two reactors was under construction in South Carolina. This project was abandoned even though the utilities had invested $9.0 billion in the project. It was less than 40% complete. The estimated cost to complete the two reactors was $25 billion compared to the original estimate of $11.5 billion.

Is there a future for nuclear power?

Increased use of nuclear power should be evaluated in comparison to the alternatives—mainly increased use of wind and solar energy. The use of wind and solar power is increasing rapidly and becoming cheaper and more reliable. They provide increasingly attractive alternatives to nuclear power. There may be a hybrid solution using both nuclear power and renewables in combination with smart electrical grids and electricity- pricing models that facilitate matching electricity supply and demand.

We should consider a plan to allow existing nuclear power plants to continue operating as long as they can do so safely. This could involve a subsidy or other incentive to ensure that electricity from nuclear plants is competitive with natural gas, at least for another 10 years or so. If these plants are shut down before the end of their service lives, much of the electricity they produce would be generated by an increased use of natural gas, with a corresponding increase in CO_2 emissions.

Before we can expand the use of nuclear power to be a major source of carbon-free energy, we need to deal with a number of practical problems.

Nuclear power plants are expensive and the electricity they produce may not be competitive with natural gas or renewables. The Vogtle power plant under construction in Georgia is already estimated to cost about $11.5 billion per reactor. According to the U.S. Energy Information Agency, the estimated cost of electricity from a new nuclear plant entering service in 2023 is $77.5 per megawatt-hour. This is the levelized cost of electricity (LCOE), a commonly used metric. This is more expensive than power generated by fossil fuel and renewable sources of energy. The LCOE estimates the present value of the cost of electricity over the lifetime of a power plant. It approximates the average price of electricity that a power plant must receive to break even over the life of the plant.

A carbon tax would increase the cost of electricity produced using fossil fuels (coal and natural gas) to make nuclear more competitive. However, the cost of wind- and solar-generated electricity is low and decreasing and will provide increasing price competition in the future.

The United States would have to rebuild its nuclear power capabilities. The U.S. nuclear industry has atrophied over the past 40 years. The engineers and other skilled people needed to staff a renewed nuclear power program have retired or are pursuing other careers. Rebuilding U.S. nuclear capabilities could only happen if nuclear power can be demonstrated to be cost-competitive in the future, and if the demand for new nuclear plants is certain enough for engineers and other skilled professionals to devote their careers to nuclear power. Companies have to be convinced that investing in the capability to finance, design, construct, and operate nuclear power plants is a reasonable business opportunity.

As mentioned above, the U.S. government needs to solve the spent fuel and radioactive waste disposal problem before the nuclear power industry could be revived. We will not know the true cost of nuclear power until we know the cost of processing and storing spent fuel rods and other nuclear waste. Today's practice of storing spent fuel rods in cooling ponds or dry casks at each reactor site is not a long-term solution.

The government should also consider recycling spent nuclear fuel as is done in France and some other countries. This greatly reduces the volume of radioactive waste that needs to be stored. It also allows the recovery of nuclear isotopes that are used in medicine, and the recovery of unused nuclear fuel that can be recycled into new fuel rods. Recycling this material into new fuel would safeguard against nuclear material being diverted for use in weapons. Problems associated with recycling spent nuclear fuel have been solved by other countries long ago.

Siting is a major problem if not handled properly. No matter what the government or others say or do, the public does not want to live near a nuclear power plant. The only solution is remote siting of nuclear facilities. Nuclear power plants produce waste heat that has to be removed by cooling water or water-cooled condensers. Some plants also use air-cooled condensers. Water cooling is preferable but uses large quantities of water. Plants would need to be located adjacent to large rivers, lakes, or the ocean. There should be several sites in the United States that would be suitable. Some sites could be on government land leased to plant owners. If remote siting was the preferred solution, there would need to be some business arrangement for utility companies to build, own, and operate nuclear power plants outside their service areas.

As stated earlier, some nuclear power plants have operated for more than their 40-year initial design lives. Some have been given permission to operate for 50 years and recently some have had their design lives extended to 80 years.

Decommissioning is another issue that clouds the nuclear power debate. What happens with old nuclear power plants at the end of their useful lives? The San Onofre Nuclear Generating Station (a two-reactor plant in Southern California) is a good example of what has to be done.[3]

This complex included two 1100-MWe nuclear power plants located on California's coast 100 km south of Los Angeles. It started producing power in 1983 and was shut down permanently in 2012 after about 29 years of operation. This was well short of its 40-year design life. The plant's steam generators had started leaking and Southern California Edison Company, the plant operator, decided to replace them at a cost of nearly $US700 million. They were replaced with units of a new design, fabricated by Mitsubishi. Unfortunately, the new steam generators had a design flaw and began leaking after less than a year of operation. Now facing a billion dollar debacle, the utility decided it was more economical to cease operations rather than undergo the substantial expense of replacing the steam generators a second time.

The actual deconstruction of these two reactors did not start until 2020, about 8 years after they were shut down. The decommissioning process is expected to be completed in 2028.

The cost of dismantling the San Onofre reactors is estimated to be $US4.4 billion compared to the initial construction cost of $US10.7 billion stated in 2018 dollars. Fortunately, the estimated cost of dismantling these reactors was accumulated in a decommissioning trust fund while the plants were operating.

---

[3] Rob Nikolewski, "The San Onofre Nuclear Plant will be Dismantled Starting Next Month," *The San Diego Union-Times*, January 26, 2020.

Huge quantities of steel, concrete, and other materials have to be disposed of. Low-level nuclear waste will be sent about 1,111 km (760 miles) by rail to a disposal site in the Utah desert. Higher-level nuclear waste will be sent about 2,250 km (1400 miles) to a storage facility in west Texas. Nonradioactive waste will be sent to a disposal site in Arizona.

What about the spent reactor fuel? All the nuclear fuel used by these two reactors, about 1800 tons, is still on site. The U.S. government has yet to come up with a solution for the reprocessing and permanent storage of spent reactor fuel. The fuel on site is being transferred from "wet storage" in large pools on site to "dry storage" in 50-ton canisters that will be stored at the decommissioned site forever or until a better spent fuel disposal plan is agreed to.

To reduce the cost of nuclear plants there need to be some economies of scale. A good idea would be to build several nuclear reactors of the same design, at the same location, using the same contractors. This would also allow cost savings associated with some shared facilities and equipment such as for refueling.

Financing could be a problem due to economic fundamentals. To be economical, a new nuclear plant would have to operate profitably for 40 years or more to pay back construction loans and provide an acceptable return on investment. Several nuclear power plants in the United States are being shut down before the end of their service lives because they are not competitive with natural gas-fueled power plants. Lower cost solar- and wind-produced electricity would be a problem in the future. The cost of solar- and wind-produced electricity continues to decline due to growing economies of scale and improving technology. Long-term power purchasing agreements or some other guarantee that the plant would be operated after it is completed would be necessary. However, what customer would agree to a 20- or 30-year contract for electricity, knowing that cheaper alternatives could be available in the future?

Insurance is a problem. The current generation of nuclear power plants needed the federal government's Price-Anderson Nuclear Industries Indemnity Act to satisfy liability claims for injuries and property damage that would result from a serious, even if very unlikely, nuclear accident. This or some similar government guarantee would have to be extended to new nuclear power plants.

New nuclear power plant designs might not evolve in time to help reduce greenhouse gas emissions. There are some promising new designs

Some successes and failures 195

for nuclear power plants that may prove to be more economical and inherently safer than current designs. However, even if one of these new designs is pursued, it will take 20 years or more before these new designs can be scaled up to utility-size nuclear power plants and built in large numbers. It could take a long time to satisfy the Nuclear Regulatory Commission safety requirements and get approval for any completely new reactor design.

Ethanol

After the Clean Air Act amendments of 1990, a chemical compound called methyl tertiary-butyl ether (MBTE) was added to gasoline as an oxygenate to reduce unburned hydrocarbons and carbon monoxide in automobile exhausts. Gasoline contained from 10% to 15% MTBE. Because of MTBE's potential carcinogenic effects, it was replaced by ethanol, basically alcohol.

The government promoted the use of domestically produced ethanol by tax credits, import tariffs, mandates, and other incentives. The Renewable Fuel Standard mandated that billions of gallons of ethanol be blended with gasoline and diesel fuel each year. Then the Energy Policy Act of 2005 mandated that increasing amounts of renewable fuels be mixed into America's fuel supplies over time. The Energy Independence and Security Act of 2007 greatly increased the mandated quantities. Today, most gasoline sold in the United States is 10% ethanol.

Demand for ethanol and ethanol prices would probably decline if subsidies and mandates were removed. Naturally, those who benefit from ethanol's subsidies and mandates do not want them discontinued.

Is ethanol a good substitute for gasoline and diesel fuel? No, for several reasons. Ethanol is not a substitute because it contains less energy than these fuels and lowers gas mileage. Currently, about 40% of the U.S. corn crop is used to produce ethanol. This added demand for corn has raised the price of corn for food production. Ethanol production competes with land for food production. Finally, as stated earlier, fossil fuels are needed to produce ethanol—to operate farm equipment, make fertilizer, and process corn into ethanol.

The United States is the world's largest ethanol producer (about 58%), followed by Brazil (26%). Brazil uses ethanol as an alternative to gasoline and most vehicles sold in Brazil are flex-fuel and can run on either ethanol or gasoline. However, Brazil's ethanol is produced from sugar cane, a much more efficient raw material for the production of ethanol. Also, Brazil has more rain and a better growing climate.

Could we produce enough ethanol today to replace gasoline? U.S. corn production uses 90 million acres of farmland and about 36 million acres of land (40%) were used to produce 15.8 billion gallons of ethanol in 2017. That is about 440 gallons of ethanol per acre. The United States consumed 143 billion gallons of gasoline in 2017. If we consumed ethanol instead, how much would we need? Gasoline contains about one-third more energy than an equivalent volume of ethanol. So, it would require 1.5 times more ethanol to replace a given amount of gasoline, or 215 billion gallons of ethanol (143 billion × 3/2). At 440 gallons per acre, it would then require about 490 million acres of farmland to grow corn to make enough ethanol to replace gasoline in the United States. This is about twice the amount of farmland in the United States today, 254 million acres, almost three times the land area of Texas, the largest state in the contiguous United States.

Even then, it would take up to 0.8 gallons of fossil fuel to produce each gallon of ethanol or a total of (215 × 0.8) = 172 million gallons. This is more than all the gasoline it would replace (Photo 13.1).

Photo 13.1 Great Plains cornfield.

Some successes and failures   197

With population growth, we will need more food. Should we divert farmland to grow inefficient biofuels, mainly ethanol? Does this make sense, especially if we have to also consume large quantities of fossil fuels to produce the ethanol in the first place? This is obviously not a good idea.

Perhaps next-generation biofuels, if they can be produced economically, will be a better fossil fuel substitute than ethanol—if they do not compete with food crops and use less energy to produce.

Must subsidies and mandates last forever? Should they become permanent due to political pressures and the lobbying of interest groups? A mandate is a form of subsidy. You force others to buy a product they may not want or at a price they would not normally be willing to pay. Most subsidies are said to be temporary, for example, to help a new industry get off the ground. When a new subsidy is created, perhaps we should also establish an end date or a set of conditions (a "Sunset Rule") that will trigger the end of the subsidy.

High-speed rail

The biggest infrastructure project in the United States today is a high-speed rail project, California's bullet train. The first phase of this project was to build a 520-mile line connecting San Francisco, Los Angeles, and Anaheim, to be completed in 2029. In 2008, California voters passed Proposition 1A approving a $9.9 billion bond issue for initial funding for the project. Work on a portion of the project in California's Central Valley did not start until seven years later, in January 2015, to build a 171-mile length of the system. This is the easiest segment of the system to build, in that there are not any tunnels or the need to cross major urban areas.

Ten years after voters approved it, the project is $44 billion over budget and 13 years behind schedule. The cost estimate for the bullet train has grown from an initial $32 billion to $77 billion or more. The completion date for San Francisco to Los Angeles service has been extended from 2029 to 2033. At present, sources of funds to complete the project have not been identified. The expected cost per mile of the bullet train is much higher than for similar systems in China and Europe. The project has been grossly mismanaged and may never be completed.

In contrast to the United States, China has a high-speed rail system with 18,000 miles of track that connects most cities. All this was built since

2008, the same year that California voters approved funding to start the bullet train. China now has two-thirds of the world's high-speed rail networks. China's objective is to complete a 24,000-mile network by 2025, connecting all major and minor population centers. One can argue that there are differences. China has an authoritarian form of central government that can plan national projects and override local objections. In Europe, the U.S., and many Asian countries there are environmental regulations and other measures that must be satisfied to protect public interests.

Lessons learned

Economics are important. On the surface this is obvious, but needs to be emphasized. When fossil fuel costs rise, consumption declines. Cheap renewable energy, on the other hand, stimulates growth. This is the key to increasing the use of carbon-free renewable energy. Renewable power projects need to be carefully planned and executed. Proper site selection is critical, as is good project management, experienced contractors, and oversight to ensure adherence to schedules and budgets. Public acceptance is critical. Having an informed public that understands the needs and benefits will avoid legal challenges and project delays. If these guidelines are observed, the world should experience a growing number of large, transformative, projects that will assist in reaching the goal of net zero greenhouse gas emissions.

PART III

CHAPTER 14

Action Plan: efficiency, power, transportation, and land use

Data summarized in previous chapters demonstrate that it is not possible to achieve net zero by 2050. Even though it is the Intergovernmental Panel on Climate Change's (IPCC) goal to keep global warming less than 1.5°C, it is highly likely that the earth's temperature rise will exceed 2.0°C (see Appendix 3). Nonetheless, we need to make a start by implementing practical measures that are currently feasible.

Do we need another moon shot?

If we are serious about stopping global warming, what should we do? Do we need something like the 1960s space program to deal with global warming? The United States sent a man to the moon and brought him back safely only 8 years after President Kennedy's challenge to the country. Can something similar be done with respect to global warming? Landing a man on the moon was an option; we did not have to do it. It was a challenge to be met, not a problem that had to be resolved. Dealing with global warming is a necessity.

Tackling global warming is not a moon shot in that we have most if not all the technology we need to deal with this problem. However, like the moon shot, we have to get started without knowing how we can achieve the objective of getting to net zero emissions by 2050. When President Kennedy in 1961 challenged the country to put a man on the moon by the end of the decade, we were not sure how to do it. We had to develop new technology to achieve this objective. With global warming, we know what to do and how to do it if we have the will and if we can get the global cooperation needed. The national effort mobilized to put a man on the moon could provide some useful lessons that can be used to stop global warming.

Reaching Net Zero.
DOI: https://doi.org/10.1016/B978-0-12-823366-5.00014-2

© 2020 Elsevier Inc.
All rights reserved.

201

Personally, the authors remember the national response resulting from the Soviet Union's launch of *Sputnik* in 1957. This achievement was repeated in 1961 when a Soviet cosmonaut was the first man in space. There was great concern that the Soviet Union was pulling ahead of the United States in science and technology. This was a threat to our national security as well as a blow to our national self-image.

Sputnik was a wakeup call that got the United States to do what was needed to take the lead in space exploration. The U.S. response was massive and immediate. The Defense Advanced Research Projects Agency (DARPA) was founded in 1958 and The National Aeronautics and Space Act was passed creating the National Aeronautics and Space Agency (NASA). The National Defense Education Act spent millions on technical education in the United States. This effort led to a substantial increase in funding for research and development and funding for education in science and technology. The United States trained a new generation of scientists and engineers and dramatically increased scientific research. It can be argued that this effort was a major contributor to U.S. leadership in science and technology.

There are several differences between the Apollo Program and what is needed to address global warming. The Apollo program was a U.S.-only program. We did not need an unprecedented level of global coordination and cooperation to solve this problem. The lives of citizens were unaffected by the space program. They did not have to make any sacrifices or change their lifestyles to put a man on the moon. The economy and employment were not affected. If anything, this massive government program was a major stimulus to the economy. The positive investments in science and technology benefited the economy for decades.

The degree of global cooperation and commitment required differentiates the international effort to reduce global warming from an Apollo-like program. It is more analogous to World War II (WWII), where a group of nations—the Allies—forged an alliance to combat the Axis powers and which eventually gave rise to the United Nations Organization.

The challenges of a global approach

In contrast, addressing global warming involves decreasing and eventually eliminating the use of fossil fuels. These are large businesses globally and employ large numbers of people. However, this transition takes place over 30 or more years with time to adjust. There would be many job-creating opportunities associated with the transition to a more energy-efficient economy driven by electricity from renewable sources, along with investments in new science and technology. Lifestyles could change in a positive manner if low-cost solar and wind power became available in underdeveloped areas and countries through distributed generation.

An Action Plan, assuming we cannot get to net zero by 2050

Even if we cannot see today how to close the gap to get to net zero, we have to start reducing greenhouse gases immediately and as fast as possible. We will have time to explore additional actions and make more realistic estimates of when net zero could be achieved. For now, we describe what we consider a realistic Action Plan to reduce CO_{2eq} emissions, knowing that achieving net zero by 2050 is impossible, given the difficulties that exist (see Appendix 3).

The biggest sources of greenhouse gases are coal used to generate electricity and oil derivatives used as transportation fuels. Coal and oil are also the world's largest sources of air pollution.

- *Base case: Business as usual.*

We estimated population growth using generally accepted projections and assume that the global economy continues to grow at current rates and that current trends in reducing energy used per unit of gross domestic product (GDP) remain constant. We recognize that recessions and other unforeseen events could alter this base case.

To estimate greenhouse gas reductions in 2050, we projected emissions each year from 2018 to 2050 assuming some progress will be made based upon present trends in increased use of renewable energy, greater use of electric vehicles, growth in nuclear generation in a number of countries, and ongoing improvements in energy efficiency. These positive changes will reduce greenhouse gas emissions, but will not be enough to reach net zero by 2050.

There will be offsetting trends. Unless disease, food shortages, or limits on human reproduction have an effect, the world's population will continue to grow, reaching an estimated 9.9 billion people by 2050, an increase of 2.3 billion people. To the extent that higher crop yields cannot be obtained, this will require more farmland for food production as well as more energy. In addition, living standards are improving in many areas of the world. Higher living standards result in higher per capita energy use as more air conditioners, automobiles, and other energy-consuming products and services are purchased.

We assume that global GDP, barring recessions or other disasters, will increase by 3.25%/year, a reasonable "business as usual" scenario. In this case, global GDP will grow from \$84.8 to \$236 trillion in 2050. Energy efficiency (units of GDP per unit of energy) is improving about 2.0%/year so that energy demand is likely to grow by about 1.25%/year between today and 2050. By 2050, global energy use will increase from 14.3 billion toe (600 billion GJ) to 21.3 billion toe (891 billion GJ) and global CO_{2eq} emissions will grow from 55 to about 81.8 $GmtCO_{2eq}$/year. This is roughly consistent with the IPCC's forecast for a low global response to global warming (see Fig. 9.2).

CO_2 emissions actually increased an alarming 2.7% in 2018, up from a 1.6% rise in 2017. We make the assumption (based on energy efficiency improvements) that future growth will be only 1.25%/year. If the trend in CO_2 emission increases is more like those of 2017 and 2018, then CO_2 emissions would be higher than our estimate.

If it turns out that the individual countries' reduction pledges included in the Paris Agreement are honored by most but not all countries, the emissions estimate listed above could be less.

Actions needed to reduce greenhouse gases, in approximate descending order of priority are as follows (Textbox 14.1).

> **TEXTBOX 14.1**
> *Important note*: Our estimates assume that it will take 5 years before improvements in efficiency begin, *except* in the case of electric vehicles and conversion of fossil fuel electric generation plants to solar/wind, as these conversions are already underway.

- *Action 1: Continue to increase energy efficiency and conservation.*

  Energy conservation and efficiency can improve from the recent 2.0%/year to 3.0%/year. The additional 1.0% increase will further reduce energy demand by 23% by 2050. Note that there are a number of indirect energy savings that result from the switch to renewables, including no energy to mine or process fossil fuels, cheaper energy transmission, compared to pipelines and tankers, and savings due to avoiding waste

Action Plan: efficiency, power, transportation, and land use 205

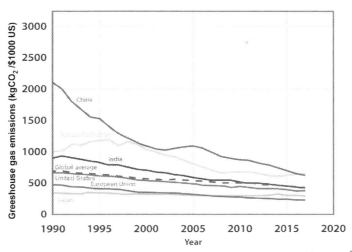

Figure 14.1 CO_2 emissions declined relative to GDP due to higher efficiency.[3]

heat when generating electricity.[1] When fossil fuels are used to generate electricity, 60%−70% of the energy input is lost as waste heat eliminated in cooling water, radiators, stacks, and exhausts. In addition, electric motors are more efficient than internal combustion engines. We believe that 1.0% is a conservative, readily achievable, estimate. Fig. 14.1 shows how CO_2 emissions have declined relative to GDP due to higher efficiency. Fig. 14.2A−C shows some specific examples of the remarkable improvements in efficiency for refrigerators, LED lights, and new building codes that have occurred since the 1973 oil embargo.[2]

---

[1] Mark Z. Jacobson, et al., "100% Clean and Renewable Wind, Water, and Sunlight All-Sector Energy Roadmaps for 139 Countries of the World," *Joule* 1 (2017): 108−121.

[2] Energy efficiency graphs. 14.2A: Marianne DiMascio, "How your Refrigerator has Kept its Cool Over 40 Years of Efficiency Improvements, *ACEEE*, September 11, 2014, pdf; Jennifer Thorne Amann, *Energy Codes for Ultra-Low-Energy Buildings: A Critical Pathway to Zero Net Energy Buildings* (Washington DC: ACEEE, 2014), Figure ES1. Source of data: U.S. DOE Building Codes Program; IEA; 14.2B: "LED Bulb Efficiency Expected to Continue to Improve as Cost Declines," March 14 2014, https://www.eia.gov/todayinenergy/detail.php?id = 15471;.14.2C: Adapted from Jennifer Thorne Amann, Energy Codes for Ultra-Low-Energy Buildings: A Critical Pathway to Zero Net Energy Buildings (Washington DC: ACEEE, 2014), Figure ES1. "Sources of historic data: International Code Council and ASHRAE. Sources of projections: Pacific Northwest National Laboratory and U.S. DOE Building Codes Program."

[3] J.G.J. Oliver, et al., "Trends in Global CO_2 and Total Greenhouse Gas Emissions, 2017 Report," PBL Netherlands Environmental Assessment Agency, https://www.pbl.nl/sites/default/files/downloads/pbl-2017-trends-in-global-co2-and-total-greenhouse-gas-emissons-2017-report_2674.pdf.

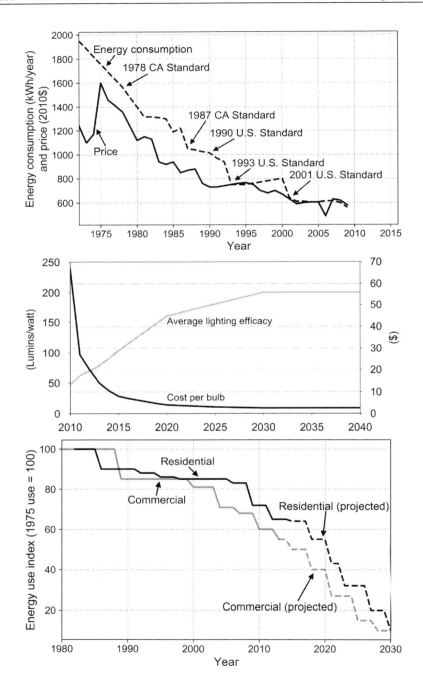

Figure 14.2 Energy efficiency graphs: (A) Refrigerators. (B) LED lamps. (C) Buildings.

- *Action 2: Stop using coal to produce electricity. Switch to renewable solar and wind in the interim where feasible.*

The world has to replace coal by solar and wind for electricity production. This is an urgent priority. In 2018, emissions from coal-fired power plants hit a new record, exceeding 10 Gmt of CO_2. Much of this came from new coal plants in Asia that have decades of operating years ahead.[4] Natural gas needs to be replaced eventually. Natural gas is a cleaner fossil fuel than coal, so replacing natural gas for electricity production is a lower priority. A growing percentage of the world's electricity is produced by renewables: hydropower, solar, and wind. Currently they account for roughly 30% of global electricity production. We need to expand efforts to improve battery storage and electrical grids to accommodate the variable nature of wind- and solar-produced electricity. By 2050, we assume that renewables can supply an estimated 50% of global electricity production. India and other Asian countries will continue to use coal to produce electricity. China will increase its use of solar, wind, and nuclear power to produce electricity but has stated that it will continue to depend on new coal plants for a significant portion of its electricity production.

- *Action 3: Switch to renewables for heat and power.*

As the price of fossil fuels rises (with or without the help of a carbon fee) and the cost of renewable energy continues to decline, more electrification will occur, increasing the demand for electricity. We estimate that an additional increase of 0.5% annually could occur, or 7744 TWh over the "business as usual" base case. As noted elsewhere, this includes an additional 3400 TWh for electric vehicles. Excluding vehicle improvements described in Action 4 (below) there would be a net increase of 4344 TWh in renewable generation, with a corresponding reduction in greenhouse gases. Residential and commercial buildings use natural gas and some oil for heating, cooking, and other needs. Most of these uses could be converted to electric power and all new construction should be designed with this in mind. Use heat pumps instead of furnaces, microwave ovens for cooking, and improve building insulation and fenestration. Industry uses oil and natural gas and some coal to generate energy needed to produce heat and power for industrial processes. Some coal, oil, and natural gas are used as a

---

[4] Chris Mooney and Brady Dennis, "Coal Plant Emissions Hit High in 2018: The Global Rise Means 'Deep Trouble' for Earth, Expert Says," *Los Angeles Times*, March 27, 2019, p. C-2.

raw material for the production of plastics, chemicals, fertilizers, and some other products. Electricity from solar and wind could displace the use of fossil fuels as a source of heat and power for most industrial applications. The production of steel and cement is the largest single source of CO_2 emissions from industrial processes. As an example of what can be done, project "Hybrit" is a feasibility study of a Swedish steel plant that uses hydrogen generated by hydroelectricity to produce steel without using fossil fuels.[5]

- *Action 4: Increase use of electric vehicles and replace petroleum-based transportation fuels.*

  About 95% of transportation uses liquid, oil-based fuels. This includes light and heavy vehicles, airplanes, ships, and railroads. The biggest potential improvements would be increasing the use of electric vehicles, followed by actions to increase the fuel efficiency of remaining vehicles. The use of electric vehicles is only a benefit if the electricity is produced by solar, wind, hydro, and nuclear energy. Some fossil fuel use in transportation cannot be displaced by renewable electricity, for example, liquid fuels used by airplanes and ocean-going ships. However, large, all-electric ferries are already operating in Scandinavia. In addition, Canada and Washington State are planning to deploy electric ferries. The number of motor vehicles is forecast to increase from about 1.2 billion today to about 3.0 billion in 2050. With population growth and improved living standards, one of the first purchases is a motor vehicle, from a scooter to an automobile. Electric vehicles will grow from 5.6 to 125 million by 2030.[6] Other estimates indicate 1 billion electric vehicles of all types by 2050, about one-third of the total. Electric vehicles will increase electricity demand and not all of this electricity will be produced by renewables. Fuel efficiency should increase so that the remaining vehicles will use less fuel per vehicle. Since vehicles last a long time, 10 or more years, it will take time to replace the current vehicle fleet with new higher mileage vehicles or electric vehicles.

[5] SSAB, "HYBRIT — A Swedish Prefeasibility Study Project for Hydrogen Based CO_2-Free Ironmaking," Presentation on May 25, 2016, https://carbonmarketwatch.org/wp-content/uploads/2016/04/SSAB-HYBRIT-A-Swedish-prefeasibility-study-project-for-hydrogen-based-CO2-free-ironmaking.pdf.

[6] Tom DiChristopher, "Electric Vehicles Will Grow from 3 Million to 125 Million by 2030." *CNBC*, May 30, 2018, data from International Energy Agency forecast, https://www.cnbc.com/2018/05/30/electric-vehicles-will-grow-from-3-million-to-125-million-by-2030-iea.html.

Action Plan: efficiency, power, transportation, and land use 209

- *Action 5: Stop deforestation and replant cleared areas.*

Deforestation is also impacting global greenhouse gas emissions by reducing carbon capture. More and more land is being cleared for agriculture, particularly to increase grazing land and to raise feed for cattle. The horrific wildfires in Australia, the Western United States, Siberia, and the Amazon region increase the impact. The earth's plant life, mainly forests, absorbs about 30% of CO_2 emissions. The destruction of forests reduces the amount of CO_2 they absorb, leaving more CO_2 in the atmosphere and oceans. This includes, for example, the clearing of tropical forests in Southeast Asia to plant palm trees to produce cooking oil and biodiesel fuel. There is a great potential for reforestation, on a scale of planting billions of trees, to reduce greenhouse gases. Planting new trees requires large land areas that have the soil, water, and climate to support new growth. Much of this land is in developing countries in Africa, South America, and Southeast Asia. An international program to pay these countries to plant and maintain new forested areas may be required. It will be necessary to monitor new forests to be sure the trees are planted and maintained. Planting new trees is a good idea but is not a substitute for preserving the mature forests we already have.

Afforestation should be a cheaper and more certain method to reduce and sequester CO_2 already in the atmosphere compared to unproven artificial methods under development that would remove CO_2 from the atmosphere using chemical processes and store captured CO_2 permanently in depleted oil wells or some other depository.

A recent study reported in *Science* magazine estimates that there is a total land area of 0.9 billion hectares (roughly the size of the United States) suitable for planting new forests. This could increase forested area by one-third without taking land from agriculture. This area could support an estimated 1.0 trillion new trees, compared to a total of about 3.0 trillion trees in the world today. This new growth could sequester up to 205 Gmt of CO_2, 25% of the CO_2 in the atmosphere today, as the trees grow to maturity over 50 or more years.[7] The countries with the most potential for afforestation are all in Africa: Rwanda, Uganda, Burundi, South Sudan, and Madagascar. These countries would, by themselves, have no incentive to undertake a major afforestation effort. In addition,

---

[7] Alex Fox, "Adding 1 billion hectares of forest could help check global warming, *Science Magazine*, July 2019. https://www.sciencemag.org/news/2019/07/adding-1-billion-hectares-forest-could-help-check-global-warming.

Photo 14.1 Subsistence farm Ethiopia.

Russia, the United States, and Australia also have major potential to restore forests (Photo 14.1).

Even though the potential for afforestation and reforestation exists, the difficulties associated with stopping the destruction of today's forests and planting billions of new trees should not be underestimated. The countries with the most potential for improvement also have very little incentive and insufficient resources to undertake the efforts required. This is another situation where a large global effort is required by the United States, China, Europe, and other developed countries to provide the resources and incentives needed before other countries will reduce deforestation, plant trees, and improve agriculture.

Action Plan: efficiency, power, transportation, and land use

- *Action 6: Improve land-use changes and agriculture by adopting "green" agricultural practices.*

  About 38% of the earth's total ice-free land area is currently used for agriculture. Urban areas cover only a small fraction of the earth's land area. Most of the agricultural land is pastureland for livestock and the remainder is for crops. Agriculture is the largest source of methane and nitrous oxides. Methane is associated with raising cattle for beef and dairy products and with growing rice. Agriculture uses large amounts of gasoline and diesel as fuel and as energy for food processing. The use of nitrogen fertilizers is the main source of nitrous oxides; organic fertilizers would help. Reducing beef and pork consumption would reduce methane emissions.

  Emissions from agriculture are likely to continue to increase by about 1.0%/year, roughly in line with population growth. Emissions due to land-use changes are assumed to remain constant as they have done for about 20 years. It is unlikely that greenhouse gases from agriculture can be reduced between today and 2050. Population growth and rising living standards will overwhelm any improvements in agriculture and also increase the pressure to clear more forests for agriculture. Also, diets change when living standards improve and people consume more meat and dairy products, requiring more land and feed for cattle.

  In agriculture, there is potential to improve farming methods to use less fertilizer and to reduce methane emissions from livestock. There is potential to store more CO_2 in soils. However, it will be difficult to make significant changes in agriculture. Except for the United States, Europe, and some other developed areas, most farming is very fragmented. It is estimated that about 1.0 billion people are employed in agriculture, most on small farms with limited access to better technology and new methods that reduce greenhouse gas emissions. Reducing deforestation also requires changes in agriculture that reduce the demand for pasture land and land to produce animal feed.

- *Action 7: Impose a carbon fee.*

  A carbon fee or tax would definitely reduce demand for fossil fuels. Already 40 countries and jurisdictions have implemented or are in the process of implementing some form of carbon fee. It is discussed in more detail at the end of this chapter. Also see Appendix 5 for more information. It would help the United States reduce fossil fuel consumption but is unlikely to be implemented in China, India, Russia, and some other major emitters. It is also unlikely that a global agreement could be reached on a universal carbon tax with penalties for those who do not comply. While we strongly support the idea of a fee on carbon, we have not included it in our Action Plan due to uncertainty regarding its prospect for global implementation on a timely basis.

- *Action 8: Accelerate research to develop effective CO_2 capture systems and systems to convert CO_2 into synthetic fuels using renewable energy.*

There is a possibility that new or more rapid technological advances could help reduce global greenhouse gas emissions by 2050. Carbon capture and storage is one such possibility, as is the greater use of hydrogen and synthetic liquid fuels produced from captured CO_2. However, use of synthetic fuels is neutral and does not reduce CO_2 in the atmosphere, as captured CO_2 is released back into the atmosphere once the fuel is burned. Theoretically, removing CO_2 from the atmosphere and storing it permanently in depleted oil wells or other repositories would reduce CO_2 buildup, but large-scale CO_2 removal is not practical today. The estimated cost of CO_2 removal is at least \$200—\$300/mt. The long-term cost forecast is \$100/mt. This is much higher than the proposed fee on carbon being discussed at the current time. There are pilot plants extracting CO_2 in limited quantities at fossil fuel power plants. There are also proposals to develop new methods of direct CO_2 capture from the atmosphere. It would be possible to produce liquid fuels from CO_2 if the cost of electricity is low enough. If a practical use could be found for CO_2 that would give it commercial value, that could partially offset the expense of carbon capture. Admittedly, these are "long-shots" but immensely valuable if they can be made practical.

- *Action 9: Should we take a second look at nuclear power?*

Not unless there is significant progress toward key problem solutions (economics, siting, and spent fuel reprocessing and storage).

For the reasons stated above, we have not included measures 6—9 in the Action Plan.

Table 14.1 summarizes those measures that in our opinion, use proven technology, are economically feasible and for which implementation is already underway. Table 14.1 is based on estimates made by the authors, derived from key parameters as published by leading authorities. Appendix 4 lists the key parameters used and the source of the data and assumptions used in the analysis.

Action Plan: efficiency, power, transportation, and land use 213

Table 14.1 Action Plan summary.

Action 1: Improve energy use efficiency

Energy efficiency is rising throughout the world with installation of more efficient LED lighting, improved heat pumps, and household appliances, building insulation, and other like measures. Energy productivity (GDP/J) is increasing so that energy growth is slowing compared to GDP. The technology and know-how exist for every country to improve its energy use efficiency by at least 1% more per year. If this is done, it will reduce CO_2 emissions from 81.8 $GmtCO_{2eq}$ in 2050 to 62.6 $GmtCO_{2eq}$, saving 19.2 $GmtCO_{2eq}$

TOTAL due to 27% savings from improved efficiency: **19.2 $GmtCO_{2eq}$**

Action 2: Convert 50% of fossil fuel electrical generation to solar and wind

Solar and wind generation is currently being installed worldwide with a current growth rate of 7.9%/year. It is feasible to replace 50% of fossil-fuel generating capacity by 2050, eliminating emissions of 8.2 $GmtCO_2$. In addition, eliminating the fuel expense of half of the fossil generating plants would save $404 billion/year, or $12 trillion over 30 years. The fuel savings would partially offset the cost of building solar and wind farms. Note: improving energy use efficiency, as noted above, also reduces the need for new generating facilities.

TOTAL due to 50% electricity generation from renewable sources:
8.2 $GmtCO_2$

Action 3: Fuel switching

Additional electrification with renewable energy of 0.5% annually will reduce fossil fuel emissions due to fuel switching by 3.58 $GmtCO_2$.

TOTAL due to fuel switching **3.58 $GmtCO_2$**

Action 4:Transportation savings

Transportation is a large consumer of liquid fossil fuels. However, there are several steps that can be taken immediately. At present there are over 1 billion cars and light trucks in the world. This number is projected to reach 3 billion by 2050. Assuming gasoline is the fuel, these vehicles will emit 8.7 $GmtCO_2$ in 2050. Hopefully the estimate of gasoline-fueled vehicles is an upper limit. The emerging trend toward autonomous cars and ridesharing vehicles may reduce the number of car owners. At the present time there are about 5.6 million electric vehicles worldwide. This number is projected to increase to 1 billion by 2050. This will reduce emissions by 2.53 $GmtCO_2$ in 2050, but will require additional electrical generation capacity of 3.4 million GWh/year, or the equivalent of 2600 additional 600 MW solar farms.

Increase electric vehicles to 900 million: 2.53 $GmtCO_2$

Improve mileage of remaining conventional vehicles to 56 mpg: 1.07 $GmtCO_2$

TOTAL transportation: **3.60 $GmtCO_2$**

Grand total, all measures: 34.63 $GmtCO_2$

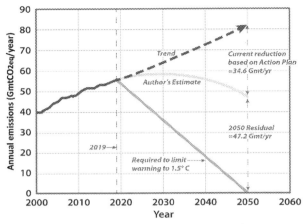

Figure 14.3 Impact of the Action Plan on "Business as Usual."

Based on Table 14.1, the "business as usual" estimate of 2050 emissions of 81.8 GmtCO$_{2eq}$ could be reduced by about 34.6 GmtCO$_{2eq}$. Emissions in 2050 would then be about 47.2 GmtCO$_{2eq}$, equivalent to the global emissions in 2010. This is an estimate based upon the assumptions listed above. This is clearly an unacceptable result, certainly not "net zero," but is probably a *realistic* estimate.

If this is the best we are likely to do by 2050, then something on the order of an additional 1672 GtCO$_{2eq}$ will be discharged into the atmosphere between now and 2050. This additional CO$_{2eq}$ would exceed the carbon budget required to stay below 2.0°C. In 2050, we would still have emissions of roughly 47.2 GtCO$_{2eq}$/year and be headed toward a 3.0°C global temperature increase. Fig. 14.3 illustrates how the Action Plan impacts the goal to reach net zero (compare to Fig. 9.5 in Chapter 9: What Would It Take to Reach Net Zero?).

Can the Intergovernmental Panel on Climate Change's goal of keeping global warming under 2°C be met?

To answer this question, we need to consider the carbon budget. The purpose of a carbon budget is to estimate how much additional CO$_{2eq}$ can be absorbed by the atmosphere before the global average temperature exceeds a specified value, nominally 1.5°C or 2.0°C. The first step in establishing a carbon budget is to estimate how much carbon is currently in the atmosphere. For any mixture of gases (such as the atmosphere) with a known concentration of any gas (such as CO_2), there is a straightforward method to calculate the mass of the constituents. Appendix 3 provides an example of the methodology.

According to the most recent IPCC Special Report (2018), 2200 ± 320 GmtCO$_{2eq}$ have been emitted from the start of the industrial revolution until about 2011.[8] The range shown for this estimate (± 320) is due to uncertainties regarding the data. About 90% of the emissions considered are due to fossil fuel use and cement production, while about 10% are due to agriculture and land-use changes. Other organizations besides the IPCC have made estimates that vary based upon assumptions used. The IPCC's estimate is for all anthropogenic (man-made) sources of CO$_{2eq}$ while some other estimates are for emissions from the energy sector only.

The next step in determining a carbon budget is to estimate what concentration of CO$_{2eq}$ in the atmosphere would cause a specified global temperature rise. This calculation is more complex and usually stated with a given probability to indicate the degree of uncertainty. The IPCC's Special Report (2018) estimated that a carbon budget for staying under 2.0°C is 1170 GmtCO$_{2eq}$ for a 66% probability of not exceeding 2.0°C.[9] (The budget would be higher if we were willing to accept only a 50/50 chance of staying under 2.0°C, for example.) To remain under 1.5°C with a 66% probability of not exceeding 1.5°C, the carbon budget is 420 GmtCO$_{2eq}$. It is important to note that the carbon budget is that amount that can be added to the 2200 GmtCO$_{2eq}$ of carbon already in the atmosphere, before exceeding the specified temperature limit. With this understanding, it is obvious that the carbon budget decreases every year, depending on how much additional CO$_{2eq}$ has been emitted annually.

Fortunately, there is an easy way to determine the current value of the carbon budget at any time. The *Mercator Research Institute on Global Commons and Climate Change* (MCC) created a "Carbon Clock" that tracks the status of IPCC's carbon budgets by subtracting the estimated daily global carbon releases.[10] According to the MCC Carbon Clock, as of August 2019, the world will exceed the IPCC's carbon budget for

---

[8] V. Masson-Delmotte, et al., *IPCCSR1.5*, 12.

[9] Ibid.

[10] The MCC carbon clock can be found on the Internet at https://www.mcc-berlin.net/en/research/co2-budget.html.

Table 14.2 Impact of Action Plan on year carbon budget is exceeded.

Carbon budget	Temperature	Year exceeded	Year exceeded
$(GmtCO_{2eq})$	(°C)	Business as usual	Action Plan
420 (IPCC)	1.5	2025	2025
1170 (IPCC)	2.0	2036	2040

1.5°C of 420 $GmtCO_{2eq}$ in only 8 years. The carbon budget for 2.0°C, 1170 Gmt CO_{2eq}, will be exceeded in 26 years if present trends continue. According to a recent United Nations report, exceeding warming of 1.5°C is *"unavoidable"* based upon present trends.[11]

Based on the Action Plan shown in Table 14.1, global CO_{2eq} emissions can be held to 47.2 $GmtCO_{2eq}$ by 2050. This is in comparison to a "business as usual model" where population and GDP continue to grow at current rates and the main mitigating factor is that the energy efficiency continues to improve by 2%/year. Cumulative global emissions to reach the IPCC temperature limits of 1.5°C and 2.0°C assuming the Action Plan is implemented are shown in Table 14.2.

This assumes a delay of 5 years before several elements of the Action Plan can take effect, except for renewable electricity generation conversions and electric vehicles, which are happening right now. It can be seen that the Action Plan outlined in Table 14.1 "holds the line" but does not solve the global warming crisis. With the Action Plan, global greenhouse gas emissions are held at 47.2 $GmtCO_{2eq}$, the 2010 level of emissions (see Fig. 9.1 in Chapter 9: What Would It Take to Reach Net Zero?). At best, the Action Plan delays the date when 2.0°C is reached by 4 years and buys time for additional developments to be implemented. It does not eliminate the possibility of a continuing rise in the average global temperature and does not eliminate the risks that will bring.

Why can't we do better?

As stated earlier, we believe the Action Plan, or something like it, is the best that can be done to reduce greenhouse gas emissions between now and 2050, given the obstacles to taking action the world faces. As

[11] UN Environment *Emissions Gap Report 2018* (Nairobi, Kenya: United Nations Environment Programme, 2018), 112 pp.

stated previously, we should start as soon as we can and do the best we can with the technology we have. If we start taking action to reduce this threat, there is a good chance that we will find ways to make faster progress toward achieving net zero. There are several reasons why we cannot do better. Here are the ones that are most important, in our opinion.

Stating the obvious, the world cannot get to net zero CO_2 emissions if the United States, the world's second largest source of greenhouse gases—and the world's largest per capita emitter—and other developed economies do not get to net zero. It is highly unlikely that the United States could essentially eliminate fossil fuel use by 2050, 30 years from now. The United States and other developed countries will still have high per capita energy uses even if these countries make additional improvements in conservation and efficiency, increase the use of renewables for electricity production to displace coal and natural gas, convert much of their vehicle fleet to electric vehicles, and make other positive changes.

As described in Chapter 11, Unique Problems of Major Contributors to Global Warming, China is leading the world in the development and implementation of renewable electricity production, the manufacture and use of electric vehicles, and the expansion of nuclear power generation. Even with these achievements, China cannot completely eliminate its use of coal to produce electricity. India and other Asian countries are heavily dependent on coal to produce the electricity they need. They are also large and growing users of oil for transportation.

A growing population will require more farmland, which is likely to lead to additional deforestation with increasing methane and nitrous oxide emissions and reduced CO_2 sequestration.

As we noted in Chapter 13, Some Success and Failures, when the 1973 embargo raised the price of oil, consumption dropped and efficiency increased. This is convincing evidence that a carbon fee or tax would reduce CO_2 emissions. However, as the 2018 French riots over a proposed increase in the price of gasoline showed, adding a carbon fee without first developing popular support will result in resistance and protests. To succeed, any such plan must be implemented gradually and should include measures to ease financial impact on the affected population that can least afford higher fossil fuel costs.

Carbon capture and storage is a big challenge. It will not be easy to develop technically practical and economically feasible systems on a scale needed to reduce CO_2 emissions by a significant amount. Carbon capture of power plant emissions may become practical if the cost of CO_2

emissions is high enough to make the cost savings from reduced emissions exceed the cost of carbon capture and storage. As stated earlier, removing enough CO_2 after it is discharged into the atmosphere to reduce the global temperature may never be practical or economical and is the most expensive solution to CO_2 reduction.

In formulating a plan, we believe it is important to set realistic goals that are *actually achievable*. There is no sense in launching an effort that has no chance of succeeding, even if it sounds good at the start. The idea that the United States could meet 100% of its power demand through clean, renewable, and zero-emission energy sources within 10 years is an example of such wishful thinking. Doing nothing based on the premise that science will find some way in the future to suck trillions of tons of greenhouse gases from the atmosphere is another example. Unrealistic plans that fail will destroy public confidence and support.

Silver bullets

For the sake of completeness we will mention some radical ideas that are categorized in general as "Geoengineering" and include concepts described as albedo modification or solar radiation management. The underlying basis for all these ideas is the observation that past major volcanic eruptions have spewed ash and sulfur into the atmosphere. For example, nearly 20 million tons of sulfur dioxide were injected into the stratosphere during Mount Pinatubo's 1991 eruptions, and dispersal of this gas cloud around the world caused global temperatures to drop temporarily (1991 through 1993) by about $1°F$ ($0.5°C$). Some examples are:[12]

- Mimicking a volcano by spraying sulfate particles high into the atmosphere.
- Spraying saltwater above the oceans to whiten low clouds and reflect sunlight.
- Thinning high cirrus clouds to allow more heat to escape Earth.
- Generating microbubbles on the ocean surface to whiten it and reflect more sunlight.

[12] Some examples of proposed albedo management techniques are found in Damian Carrington, "Reflecting Sunlight into Space has Terrifying Consequences, Say Scientists," *The Guardian*, November 26, 2014 and in David Rotman, "A Cheap and Easy Plan to Stop Global Warming," *MIT Technology Review*, February 8, 2013.

- Covering all deserts in shiny material. (Maybe with solar panels?)
- Growing shinier crops to reflect sunlight.
- Launching millions of reflective balloons.
- Scattering trillions of silicon beads into the atmosphere.

One problem with most of these concepts is that, once released into the atmosphere, there is no way to control particle distribution and no way to predict where on the globe the effects, if any, would be felt. How much energy would be required? The Mount St. Helens eruption was equivalent to 24 megatons, more than a thousand Hiroshima A-bombs. Mt. Pinatubo's energy release was 10 times greater than Mount St. Helens. What about cost? No one can say for certain what it would cost to implement one of these measures on a global basis.

We do not recommend any of these approaches without first implementing a comprehensive research and development program beginning with a proof of concept on a laboratory-scale basis.

Mitigation

Even if we get to net zero emissions eventually, there is a large and growing need to mitigate the effects of global warming resulting from past and future CO_2 emissions. We will have to deal with increasing temperatures, extreme weather events, a warmer ocean, a rising sea level, more and more severe storm surges, and other unavoidable effects. We could see crop failures and lower crop yields due to higher temperatures, changes in rainfall and drought conditions. We have already seen that these effects could result in people migrating from areas that are less habitable or that can no longer feed their populations. In Alaska there are communities such as the city of Shishmaref that are threatened by rising seas, but the Federal government has not come up with funds to relocate them to higher ground. Several indigenous Alaskan villages also applied for grants, but were turned down. The Isle De Jean Charles in Louisiana won a grant of $48 million from the U.S. Department of Housing and Urban Development (HUD) to relocate its residents to the mainland. This was intended to be a model project for coastal relocation. Due to disagreements between Louisiana and the Indian tribe that had the majority of residents on the island, the project has not gone ahead.

Coastal areas throughout the world will be early victims of global warming. As described elsewhere in this book, some South Pacific islands have been rendered uninhabitable or completely overrun by rising seas, forcing the population to evacuate. Other areas, such as Shanghai and Bangladesh, have large populations exposed to flooding that will disrupt entire towns and fishing villages. In the United States, coastal portions of Florida, the southeast United States, and the West Coast are already experiencing the early effects of rising seas.

In California, from the north coast south to San Diego, coastal cliffs are crumbling, seawalls collapsing, and waterfront homes are undermined and broken up.[13] A month after a special report was published by *The Los Angeles Times*, three people were killed when a seaside bluff on Encinitas Beach suddenly gave way, burying them. In nearby Del Mar, California, the Amtrak tracks for a major north-south railroad are within a few yards of crumbling seaside cliffs. This critical north-south rail line carries about 8 million passengers and more than $US 1 billion in freight every year.

The local agency, the San Diego Association of Governments, has spent $US 15 million over the last several decades and is now facing another $US 10 million expense for immediate repairs plus $US 100 million over the next 4 years, The alternative: spend $US 4 billion to relocate the tracks inland.[14]

The costs of moving threatened coastal communities will be substantial. Sea wall construction, just for the United States, will cost over $400 billion during the next 20 years and will only be a temporary measure (see Fig. 14.4). In addition, the impact of global warming and associated climate changes will be distributed unevenly. Some regions and some people will be affected much more than others. Those affected the most are likely to be those who are least able to withstand the effects of global warming. It is possible but unlikely that a global effort to more equitably share the cost of mitigation will be agreed to by major greenhouse gas emitters. See Appendix 5, Flood and Sea Rise Mitigation, for additional details.

These examples highlight some of the challenges raised by sea level rise. There are no easy, inexpensive, or permanent fixes. At best mitigation measures can buy time, but unless a rapid reduction of greenhouse

---

[13] Rosanna Xia, "Our vanishing coastline—California against the sea," a *Los Angeles Times* special report, July 7, 2019.

[14] Joshua E. Smith, "Del Mar ready to shore up its bluffs," *Los Angeles Times*, January 26, 2020.

Action Plan: efficiency, power, transportation, and land use   221

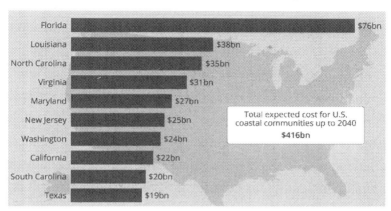

Figure 14.4 Estimated cost of seawall construction in the United States, through 2040.[15]

gas emissions takes place soon, money spent on mitigation will be completely wasted. Typical measures are as follows:
- Berms: can deflect wave action or tidal flows to help preserve beaches.
- Install riprap or grouting cliff faces: riprap or grouting can help reduce cliff erosion.
- Breakwaters: can reduce wave impacts.
- Seawalls: can prevent flooding.
- Floodgates: can be closed to protect harbors under elevated sea conditions.
- Elevate structures: require higher foundations for structures in low-lying areas.
- Gray-green living shorelines. The concept is to combine a breakwater or riprap protection with a dune and plants to stop coastal erosion.[16]
- Managed retreat: demolish endangered structures and rebuild inland.

These measures are listed in approximate order of increasing cost. They must be considered temporary in nature, as rising seas will eventually defeat them. In some cases there are temporary benefits followed by adverse effects. Berms and breakwaters can prevent beaches from receiving

---

[15] Niall McCarthy, "U.S. Expected to Pay Over $400bn on Seawalls up to 2040," *Statista*, June 21, 2019, https://www.statista.com/chart/18457/estimated-expected-cost-of-seawall-construction/. Used with permission from Niall McCarthy and *Statista*.

[16] Personal communication, Brett Sanders, UCI. See also: Kristen Goodrich, et al., "Co-development of Coastal Sediment Management Approaches in Social Ecological Systems in Southern California," *Coastal Sediments 2019*, Proceedings of the international Conference pp 1821–1825, May 2019.

seasonally placed sand, leading to their destruction. Seawalls are only beneficial to a certain point, beyond which costs become prohibitive.

Floodgates have been deployed in the Netherlands and more recently are under construction in Venice. They are very expensive. The plan to protect the Venetian lagoon with flood gates is a 20-year, $US6 billion construction project. It is designed to protect against a 60-cm sea level rise. Docks and other waterfront structures can be elevated and critical port facilities moved inland to higher ground. Managed retreat can take several forms. One example is abandoning a coastal highway and rebuilding it further inland. A more controversial form of managed retreat is to condemn and demolish waterfront homes and businesses and then rebuild them further inland. The vacant land left behind can be shaped into a beach that prevents further erosion. At issue is the question, "who pays?" If a homeowner remains in an exposed area refusing to move, who is liable? Once damage occurs somewhere, property values are likely to decline rapidly and insurance will become unavailable. Clearly, there are difficult choices and no easy answers.

Storms and flooding mitigation: Inland areas are also subject to flooding. With global warming, severity and duration of storms increase. Greater rainfall will cause rivers to swell and low-lying areas to flood. Dikes and flood control measures will fail. Mitigation measures include:

- dams
- dikes
- drainage systems
- pumping systems
- intentional flooding of low-lying areas

Much of the existing infrastructure to control flooding has been designed to withstand historic "worst conditions," typically specified as a 100-year storm or 100-year flood. The problem is that with global warming, the historic worst condition no longer applies. Storm frequency and duration are increasing. Dikes, many in use for decades, are no longer adequate for the intensity of new storms. As noted elsewhere, pumping systems to avoid flooding in Houston and New Orleans have been overwhelmed by recent storms. In the Netherlands, flood control measures include the capability to allow flooding of indefensible lands as a last-ditch resort to protect populated areas.

High temperatures that accompany global warming can be harmful to humans, livestock, agriculture, forests and wildlands. Possible mitigation measures include:

Action Plan: efficiency, power, transportation, and land use 223

- Provide shaded shelter areas.
- Increase use of evaporative cooling in suitable climates.
- Increase air-conditioning.
- Establish standards for better insulation of buildings and farm structures.
- Upgrade utility infrastructure to ensure the availability of electricity and other essential services.
- Change zoning regulations to restrict building in potential wildfire areas.
- Upgrade emergency services to respond to more frequent and more severe wildfires—warning systems, evacuation, and firefighting.

Carbon fee

Why do we need a carbon tax or fee? Everyone hates new taxes. There are at least two good reasons.

First, the fossil fuel industry benefits from huge subsidies. There are tax incentives associated with exploration and development. There are cash subsidies in many countries, mostly developing countries, under which gasoline, diesel fuel, and electricity are sold below cost with the government paying the difference. However, the biggest subsidies are associated with air pollution and the emission of greenhouse gases into the atmosphere for free. The fossil fuel industry needs to pay for the damage associated with these emissions. It should not be free.

Second, a fee would raise the cost of fossil fuels relative to renewables and provide an added incentive for customers to switch from fossil fuel use to energy from renewables, mainly solar and wind. The cost of solar and wind is decreasing relative to fossil fuels and there is already a trend toward renewables. However, the transition is not moving fast enough to address global warming. A carbon fee would accelerate this change.

What is a carbon fee? The concept is to impose a fee, in terms of a dollar amount per ton of CO_{2eq} that would be released by each ton of coal, barrel of oil, or cubic meter of natural gas produced.

"Show me the incentives and I will show you the outcome"[17] is a quote from Charlie Munger, Warren Buffet's long-time investment partner. Munger is Vice Chairman of Berkshire Hathaway, Inc., Buffet's

[17] "Industry needs more carrots than sticks," *Financial Times*, https://www.ft.com/content/ef09271c-b70f-11e2-a249-00144feabdc0.

investment firm. The choice of which fuel to use is very sensitive to relative cost. For example, natural gas quickly displaced coal in the United States as lower priced natural gas became available due to fracking. If we get the incentives right, solar and wind power should increasingly displace fossil fuels without government subsidies, mandates, and other regulations. We can speed up or retard this process depending upon the policies and practices we adopt. Increasing the cost differential between electricity generated from solar and wind relative to electricity generated by burning fossil fuels will accelerate the rate at which renewables replace fossil fuels.

As wind- and solar-generated electricity become cheaper, more reliable, and we are better able to match supply and demand for electricity from these sources, electric utilities and their customers will naturally switch to renewables as long as the declining price of renewable energy is passed on to consumers.

If lower cost electricity from renewables lowers retail electric bills, this will help displace gasoline and diesel powered vehicles with electric vehicles. This would happen for a wide range of consumer and industrial applications, from heat pumps and electric hot water heaters to the use of electricity for process energy in manufacturing. Electricity, by its nature, is a much cleaner form of energy than fossil fuels, so long as it is generated from renewable sources. It is much easier to transport and distribute to end users. This is an inherent advantage of electricity.

What should be done? Subsidize renewable energy, solar and wind, or increase the cost of fossil fuels with a tax or carbon fee on oil, gas, and coal? There is strong public opposition to taxes that increase the cost of fossil fuels, especially on gasoline and diesel fuel. Perhaps we should ask the same question in a different way? Should fossil fuel users be able to discharge CO_2 and other pollutants into the atmosphere at little or no cost? Asked this way, it is likely that more people would be in favor of placing a fee on emissions from fossil fuel use.

Any carbon fee or surcharge would have to be applied gradually to allow economies and individuals to adjust to higher cost of fossil fuels. In forecasting energy use and greenhouse gas emissions, we think it likely that a carbon fee will be implemented to offset fossil fuel subsidies and provide additional incentives to switch from fossil fuels to renewables. We anticipate that any implementation of a carbon fee will be uneven with a number of problems. There is a risk that some countries will take much longer than others to implement a carbon fee, or fail to implement a carbon fee at all. There will be inconsistencies with the fees imposed and

Table 14.3 Effect of $US50/ton of CO_2 carbon fee on fuels.

Fuel	Price increase (with $50/ton of CO_2 equivalent)	
	Amount ($US)	Percent (%)
Gasoline (gallons)	0.44	22
Natural gas (1000 cubic feet)	2.66	62
Home heating oil (gallon)	0.29	28
Coal (short ton)	105	330

enforcement efforts. The best approach would be to apply a fee or surcharge at the source—the well head or mine.

We recommend that a fee be applied to fossil fuels when they are produced or imported instead of a tax on emissions or a Cap and Trade scheme. A simple fee on production or imports would be easier to collect, apply equally to all users, and be less vulnerable to gaming by politicians and special interests who can carve out special treatment with less visibility when the rules are complicated. Taxing emissions or Cap and Trade requires a much larger organization and more complex regulations to identify, tax, and value all emissions for trade purposes. A fee or tariff should be applied on imports from countries that do not implement a carbon fee themselves.

For the United States, an assumed fee of $50/ton of carbon dioxide emissions would have produced gross receipts of about $US200 billion in 2018. If this fee is increased 2.0%/year, the gross receipts for 10 years, 2018—2027, would total $US2.1 trillion. These results are from the E3 Carbon Tax Calculator.[18] This calculator only applies the fee to fossil fuels used as fuel and excludes non-CO_2 greenhouse gas emissions such as those from agriculture or land-use changes. This calculation takes into account the likely reduction in fossil fuel use after the tax is introduced.

The estimated impact a $50/ton carbon fee would have on the retail prices of commonly used fuels as shown in Table 14.3. One benchmark to price CO_2 emissions would be to charge emissions the same price as the cost of removing CO_2 from the atmosphere. As mentioned earlier, this could be hundreds of dollars per metric ton.

---

[18] Marc Hafstead, "Introducing the E3 Carbon Tax Calculator: Estimating Future CO_2 Emissions and Revenues," *RFF Common Resources*, September 25, 2017, https://www.resourcesmag.org/common-resources/introducing-the-e3-carbon-tax-calculator-estimating-future-co2-emissions-and-revenues/.

Rather than tax fossil fuels, would not it be easier to subsidize renewable sources? Subsidies are complex to administer and are open to granting special treatment to favored organizations or special interests. Where would the money come from to subsidize renewable energy sources? If these costs are passed on to consumers as they are in Germany, it will increase the cost of electricity from renewable sources. It also reduces incentives to lower the cost of renewables through greater investments and new technology. The best solution is to pass on the declining cost of solar- and wind-generated electricity to consumers. Texas is a good example of how this could be done. As noted in Chapter 11, Unique Problems of Major Contributors to Global Warming, in Texas, electricity suppliers compete for customers' business. Electricity prices have declined due to competition, while they have risen in California. Appendix 5 describes Cap and Trade and other types of financial measures proposed to slow global warming.

CHAPTER 15

Can it be done?

Can renewable energy power the world? It appears technically feasible. A number of people and publications have denied the feasibility of transitioning to renewable energy sources, by adding solar and wind to the existing base of hydropower and nuclear, over the next 30 years or so. Yes, there are problems to be solved. It will be expensive, but a significant part can be paid for by redirecting a portion of the money spent on fossil fuels as the use of fossil fuels declines. We have access to the technology and capital required to make the transition if there is public support and political will.

Can we afford not to? Many people and organizations that object to the cost and changes needed to address this problem deny or minimize the future effects of global warming. It is analogous to a patient ignoring cancer and objecting to the recommended surgery, radiation, and chemotherapy needed to save the patient's life.

Let us look at the feasibility of providing the energy we need from renewable sources, solar and wind. Is it even possible to supply enough energy from them?

The trend is our friend

The cost of electricity generated by solar panels and wind turbines has dropped dramatically over the past 20 years. Future decreases can be anticipated due to technical improvements and economies of scale as the number of solar and wind installations increases. In this sense, the trend (cost) is our friend. Electricity from solar and wind is increasingly cost-competitive compared to fossil fuels and nuclear. The use of fossil fuels to produce electricity is a mature technology and unlikely to show significant improvements relative to solar and wind (Photo 15.1).

We obviously have to be sure that electricity from renewable sources is as reliable and available as that from fossil fuel power plants. As stated earlier, a big problem to be solved is to accommodate the variable output

Reaching Net Zero.
DOI: https://doi.org/10.1016/B978-0-12-823366-5.00015-4

Photo 15.1 Solar and wind farms, North Palm Springs, California.

of solar and wind power plants to match supply and demand. Considerable effort is being directed at solving the problem of economical, large capacity, energy storage systems and smart electrical grids for this purpose.

Fig. 15.1 is from a recent study by a major investment bank comparing the levelized cost of energy (LCOE) alternatives. It clearly shows the significant decrease in energy costs of solar and wind compared to fossil fuel alternatives. The LCOE for various sources of electricity is a fair method of comparing the cost of different energy technologies. It is the total life cycle cost of electricity for a given technology divided by the total life cycle electricity produced, expressed as dollars per megawatt hours ($/MWh). Levelized cost includes the capital costs, fuel costs, operation and maintenance, financing, and assumed utilization rate. However, these data do not include the cost of storage that would be needed for solar and wind.

The choice of energy source by consumers largely depends on the relative cost of the alternative sources. This is demonstrated by the rapid switch from coal to natural gas in the United States, as natural gas prices declined significantly when supplies increased due to fracking. If electricity from solar and wind is less expensive than fossil fuel-generated electricity, market forces will cause energy customers to switch to renewable energy

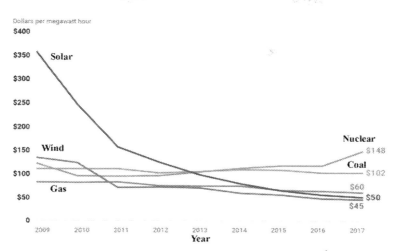

Figure 15.1 Levelized cost of energy alternatives in North America.[1]

over time. This assumes the decreasing cost of electricity from solar and wind is passed on to consumers. If government regulations, price setting by regulators, or other factors prevent cost decreases from benefiting customers, it will retard the switch from fossil fuels to renewables. In addition, fossil fuels are heavily subsidized, which reduces the economic incentive to transition to renewable energy sources. We discuss this in more detail under "Can we afford it," later in this chapter.

Oil, natural gas and coal have depletion allowances that range from 10% (coal) to 15% (shale oil). The depletion allowance acts as a tax credit. In the case of coal, for example, 10% of a coal mine's annual revenue is deducted from income taxes. There are many other forms of government subsidies, direct and indirect. Internationally, oil-producing countries spend between $775 billion and $1 trillion annually to subsidize fossil fuels, while the United States spends $20 billion annually.[2] This expenditure could be redirected to financing solar and wind projects or the required infrastructure upgrades.

---

[1] Lazard,'s Levelized Cost of Energy Analysis—Version 12.0, November 2018, https://www.lazard.com/media/450784/lazards-levelized-cost-of-energy-version-120-vfinal.pdf. Used with permission from Judi Mackey, Managing Director, Global Communications, Lazard.

[2] Oil Change International, "Fossil Fuel Subsidies Overview," accessed February 9, 2020, http://priceofoil.org/fossil-fuel-subsidies/.

Can renewable energy power the world?

Recently detailed and comprehensive studies have been published in Europe by Lappeenranta University of Technology[3] and in the United States by Stanford University.[4] They confirm that a 100% renewable electricity system is technically and economically feasible worldwide.

Global energy use in 2018 is estimated to be about 14.3 Bmtoe/year (billion metric tons of oil equivalent per year), or 166 trillion kWh/year.[5] This is not just electricity use. It is total energy use, assuming that all this energy was consumed as electricity. This approximates what needs to be done if we are to eliminate fossil fuels. Can renewable sources of energy, mainly solar and wind, produce this much electricity to power the global economy?

In reality, we do not have to replace all this energy, even though we assume 100% replacement in this analysis. Most fossil fuels are burned in a power plant or engine to produce heat that is converted into useable electricity or mechanical power. As stated elsewhere, this energy conversion process is only 30%–40% efficient. Sixty to seventy percent of the energy in fossil fuel is eliminated as waste heat.

Solar energy has a low power density in that it takes a large area to collect enough energy from solar radiation to produce electricity on a utility scale. This is compounded by the fact that solar panels only produce power about 5–6 hours per day when the sun is shining brightly. The amount of sunlight varies by time of day, month of the year, and location. Average daily output also varies depending on local weather conditions. The following is a simple illustration of how to determine the amount of electricity that can be produced by solar arrays. Obviously, a solar array in the Nevada desert or the Sahara Desert will produce more power than an array in Berlin or Boston.

---

[3] Energy Watch Group "New Study: Global Energy System Based on 100% Renewable Energy," April 12, 2019, http://energywatchgroup.org/new-study-global-energy-system-based-100-renewable-energy.

[4] Mark Jacobson, "100% Clean and Renewable Wind, Water, and Sunlight All-Sector Energy Roadmaps for 139 Countries of the World," September 2017, https://web.stanford.edu/group/efmh/jacobson/Articles/I/CountriesWWS.pdf.

[5] Global energy consumption in 2018 increased at twice the average rate since 2010, 2.3 percent/year to 14,301 Mtoe. See *World Energy Outlook 2018* (Paris, France: IEA, 2018), https://www.iea.org/weo2018/.

Can it be done?

Table 15.1 Electrical output from 1 m^2 solar panel.

Sunlight (kW/m^2)	Solar panel efficiency (%)	Days/ year	Hours of sun/day	Total annual output (kWh/m^2/year)
1.0	25	365	5	456

For the purposes of our simple analysis, we use an estimate of 1000 W of sunlight as the amount of the sun's energy that falls on a one square meter solar panel. This is a common industry estimate. One thousand watts is one kilowatt. To get the average energy output of this typical solar panel, we need to take into account the conversion efficiency of the solar panel, the number of hours per day during which the sunlight is intense enough to produce power, and the number of hours in a year. Table 15.1 illustrates how solar power can be estimated.

If we divide the earth's total energy use in 2018 (166 trillion kWh/ year) by the output of our typical solar panel, we determine the total area of panels—364 billion m^2—needed to power the earth with solar energy. Dividing by 2,590,000 m^2/mi^2, we find that an estimated 140,000 square miles of solar panels are needed to power the earth. This is equivalent to a square 374 miles on a side. Obviously, all these solar panels would not be in one location but would be distributed on rooftops and in solar farms worldwide.

Regarding land area, the earth has 57.3 million square miles of land, of which 33% is desert and 24% is mountains. The area required for this number of panels is equivalent to 0.7% of the world's deserts. Powering the globe with renewable electricity would take about 4% of the Sahara Desert, for example. Realistically, the land area would be somewhat greater, depending on how much was rooftop installations versus solar farms. Solar farms require some additional land area for access aisles, roads, and auxiliary equipment.

There would also be the investment required for the associated transmission and distribution system and other infrastructure costs. We would have to deal with the difficulty of matching the intermittent supply of energy from renewable sources with electricity demand. This problem is receiving attention and will be solved eventually. We will need some storage capacity to smooth out differences between supply and demand, large capacity batteries plus a smarter distribution network to better match supply and demand in real time, and flexible time of day pricing to give customers an incentive to switch part of their electricity demand to times when electricity is cheapest. For example, consumers could charge electric vehicles during the day at work rather than at home at night.

Can wind power the world?

Wind could also power the earth. A typical onshore wind turbine has a rated output of 4.0 megawatts. Much larger and more efficient wind turbines are being installed today. Offshore wind turbines have higher wind speeds and steadier winds. However, they are more expensive to install. For our analysis, we will use 4.0-megawatt onshore wind turbines, a conservative estimate. See Table 15.2 for the electrical output from a 4 MW wind turbine.

The wind does not blow hard enough all the time (over 20 miles/hour) to produce the rated capacity continuously, so rated capacity has to be multiplied by a capacity factor, typically 40%, to get the actual average output of a wind turbine. Average output can be further reduced if the wind turbine is within a wind farm where some of the wind energy is absorbed by other upstream wind turbines.

If we divide the earth's total energy use of 166 trillion kWh/year by the output of our typical wind turbine, we determine that it would take 12 million 4.0-megawatt wind turbines to power the world. This is equivalent to 56,000 wind farms, sized at 850 MW, each with 213 such turbines. In 2018 there are about 2000 wind farms in the world, with China, the European Union, and the United States having the most.

Wind farms today are being constructed with generating capacities of 600 to more than 1000 MW. Depending on location, the area required is comparable to a solar farm. Wind turbines located alone or along ridgelines require less land than those located in large wind farms. In large wind farms the turbines have to be distributed so that adjacent turbines do not interfere with the amount of wind other turbines receive. The area

Table 15.2 Electrical output from 4 MW wind turbine.

Rated capacity (MW)	Capacity factor (%)	Hours/ year	Annual output (megawatt-hours)
4.0	40	8760	14,000

within a wind farm can be used for other purposes such as for farming, for industrial facilities and warehouses, or even for solar panels.

An average 850 MW wind plant requires an area of 165 km^2, so as an upper limit these wind farms would require an area equal to about 18% of the world's deserts. However, by selecting optimum sites including off-shore sites, the total land area would be reduced.

What would it cost?

A growing number of reputable sources support the finding that the cost of wind- and solar-generated electricity is dropping rapidly and is now cheaper than coal and on par with natural gas in many locations, including China and India[6],[7],[8] (see Table 15.3). In comparison to the costs in Table 15.3, the typical cost for new hydro plants is $5000/kW, and for coal plants, $3500/kW. Nuclear plants in the United States cost at least $5000/kW. China is quoting $2500/kW for nuclear plants. Only natural gas is currently competitive with renewable energy at $1000−$2000/kW, but has a higher cost of generation due to fuel cost.

Table 15.3 Solar and wind power costs, including storage.

Plant type	Typical size (MW)	Cost[a] ($/kW)	Cost to power the earth ($ trillion)
Solar	600	1000	76
Wind	850	1000	47
Storage	30−400	150[b]	Not available

[a]Costs are dropping rapidly with newer plants forecast to be $900−$1000/kW.
[b]For 60 MW storage capacity.

[6] Phil Dzikiy, "Renewable Energy Costs Hit New Lows, Now Cheapest New Power Option for Most of World," *Electrek*, May 29, 2019, https://electrek.co/2019/05/29/renewable-costs-new-lows-cheapest/.

[7] Phil Dzikiy, "New Wind and Solar Now Cheaper than 74% of Existing US Coal Plants, Study Says," *Electrek*, March 27, 2019, https://electrek.co/2019/03/27/wind-solar-cheaper-coal/.

[8] Harold Anuta, Pablo Ralon, and Michael Taylor, *Renewable Power Generation Costs in 2018* (Abu Dhabi: International Renewable Energy Agency, 2019), 88 pp., https://www.irena.org/publications/2019/May/Renewable-power-generation-costs-in-2018.

Remember, solar and wind have no fuel cost. The added cost of storage is expected to add 20%–30% to the cost of solar or wind electricity.[9]

Can we afford it?

Can we afford not to? It should be clear that we cannot continue to rely on fossil fuels for our energy needs and continue discharging huge quantities of greenhouse gases and other pollutants into the atmosphere indefinitely. We will have to make a transition to renewable sources of energy sooner or later.

As noted earlier in this chapter, the cost of electricity from solar and wind is low and decreasing, while the cost of electricity produced using fossil fuels, a mature technology, is not getting any less expensive. In an increasing number of locations, electricity from renewables is cheaper than from fossil fuels and nuclear power. This trend will continue due to technology improvements and economies of scale resulting from the increasing use of solar and wind to generate electricity as shown in Fig. 10.4A and B (Chapter 10: Energy Alternatives).

Even with the decreasing cost of electricity from renewables, we have to consider the investment required to make the transition from fossil fuels to renewables. Global *total* energy use (not just electricity) in 2018 is estimated to be 166 trillion kWh/year (599 billion GJ/year). Noting that one watt is one Joule/second, the hypothetically equivalent generating capacity would be 19 million MW. If we assume that the installed cost per kilowatt of wind or solar plants, or some combination of the two, averages about $1000/kW ($1 million/MW), then hypothetically powering the world with electricity from renewable energy sources would take an investment of about $19 trillion. We would also have to add the cost of energy storage, smart electrical grids, and transmission lines, at possibly 50% more, or say $9 trillion, for a total of $28 trillion. These are very big investments but they would be spread over 30 years or more, about $1 trillion per year. This is from one-third to one-fifth the cost of an equivalent amount of natural gas, coal, or nuclear generating capacity.

[9] John Weaver, "Utility Scale Solar Power Plus Lithium Ion Storage Cost Breakdown," *PV Magazine*, January 2, 2019, https://pv-magazine-usa.com/2019/01/02/utility-scale-solar-power-plus-lithium-ion-storage-cost-breakdown/.

It is not unreasonable to conclude that there should be funds available to pay for this transition from fossil fuels to renewables over the next 30 or more years. Yes, there will be problems. For one, the benefits associated with a transition to renewable energy do not necessarily go to those who bear most of the cost. Most of the benefits from a substantial reduction in air pollution and the reduced impact of global warming are widely distributed to the global population while the associated costs have to be paid by specific individuals and organizations.

Some of the financing required will come from ongoing energy investments that are being redirected from fossil fuels plants and related infrastructure to investments in renewables. This is already happening. The elimination of subsidies for fossil fuels is another source for funding along with potential revenues from a carbon fee. In addition, there are cost savings associated with reducing the damage caused by air pollution associated with fossil fuel use, plus the reduced costs associated with mitigating the effects of global warming. Renewables are fuel-free so that today's fossil fuel costs are a future saving.

Two recent studies provide useful information. The International Energy Agency (IEA), part of the United Nation's Organization for Economic Cooperation and Development, publishes an annual report on world energy investments. From the latest report, *World Energy Investment 2019*, we can determine that the world's current investment in all forms of energy, including distribution, totaled $1.8 trillion in 2018, about 2.3% of global gross domestic product (GDP). Between now and 2050 an estimated $110 trillion would be invested in energy, assuming a 3.25% per year increase in the real growth rate of GDP during that period (the same growth rate used in Chapter 14: Action Plan: Efficiency, Power Generation, Transportation, and Land Use). This is a constant dollar estimate that excludes inflationary increases. Over time, much of this annual investment could be redirected from investments in fossil fuels to renewables.

The International Monetary Fund (IMF) recently updated its study of fossil fuel subsidies.[10] According to this report, fossil fuel energy subsidies are estimated to be $US5.2 trillion, about 6.4% of GDP in 2017

---

[10] David Coady, Ian Parry, Nghia-Piotr Le, and Baoping Shang, "Global Fossil Fuel Subsidies Remain Large: An Update Based on Country-Level Estimates," *International Monetary Fund*, May 2, 2019.

Table 15.4 Estimated fossil fuel subsidies.

Subsidy	Amount ($ trillions)	Percentage of total subsidies (%)	Percentage of global GDP (%)
Air pollution	2.5	48	3.1
Global warming	1.3	24	1.6
Underpricing fossil fuels	0.7	15	0.9
Environmental cost of road fuels	0.7	15	0.9
Total	5.2	100	6.5

(see Table 15.4). Global GDP was $US81.0 trillion in 2017 according to the World Bank. Those subsidies include the dollar cost of subsidizing fossil fuels and the cost of electricity. The biggest subsidies are the annual cost of pollution related to fossil fuel use plus the annual cost of the effects of global warming, both of which are increasing. The report includes a separate estimate of the environmental cost of road fuels. This includes the cost of traffic congestion and traffic accidents.

In making any global estimates, there are obvious disagreements on methodology and assumptions that would result in different estimates. However, we can agree that on a global scale, the amounts are huge and these two reports are probably the best, unbiased estimates we can get.

With the right incentives, a high share of the capital required will come from private sector investors making intelligent investment decisions in profitable growth businesses. This is already happening, but not fast enough. If we get the incentives wrong, private sector investment would be reduced. A greater share of the investment required would have to come from government spending plus mandates and regulations that try to push the energy industry and energy consumers toward renewable energy sources. Getting the incentives right is preferable to subsidies and mandates.

So, roughly, how would we pay for the transition from fossil fuels to renewables?

Power plants typically have a design life of 40 years. Over the next 30 years, it is reasonable to assume that most of the power plants in operation today will need to be replaced. Most could be replaced by solar and wind instead of building another fossil-fueled or nuclear power plant. Based on recent estimates, it would be cheaper to replace fossil fuel plants with solar and wind in most locations. This is starting to happen (Table 15.4).

Solar and wind installations are fuel-free. The fuel cost savings associated with renewable energy should be substantial. For example, in 2018, the world used an estimated 26,700 TWh of electricity. A good estimate of the average fuel cost per kilowatt-hour is $0.025. If all this electricity was produced by solar and wind, the annual fuel savings would be $667 billion/year.

There is a large and rising cost associated with the damage done by global warming. A big unknown is how to estimate the cost of doing nothing and having to live with the increasing costs and problems associated with global warming. These costs and other problems are frequently ignored by global warming critics and deniers. According to the IMF, the annual cost is $1.3 trillion and growing.

The elimination of most air pollution would be worth a lot. Air pollution is largely associated with using coal to generate electricity and oil-based transportation fuels. This would lead to lower health care costs, for example, and improved quality of life. According to the IMF, the annual cost is $2.5 trillion.

Carbon fee revenues could be substantial. In 2018, estimated CO_2 emissions from fossil fuels were 37.1 Gmt/year (Fig. 6.4, Chapter 6: How Do We Know Man-Made CO_2 Is the Issue?). This is the equivalent to 10 Gmt/year of carbon only. If there was a carbon fee of about $40 per metric ton, the annual carbon fee revenue would be about $400 billion. This fee would increase over time.

Today, fossil fuels are heavily subsidized. These subsidies could be eliminated or redirected to support renewable energy. Eliminating fossil fuel subsidies would increase government revenue and reduce expenses by an estimated $3.0 trillion or 3.8% of global GDP annually. Politically, it would be very difficult to reduce or eliminate fuel subsidies that the general population is used to, especially in lower-income countries.

Because electricity generated from wind and solar is fuel-free, the need for very expensive exploration that is required to find new oil, gas, and coal sources would be eliminated. The need to develop and maintain oil wells and mines is eliminated as well as the infrastructure for transporting and refining fossil fuels: oil tankers, pipelines, oil refineries and the means for local fuel delivery. Oil spills and other problems with fossil fuels would be reduced or eliminated.

Many countries such as Japan, China, India, and South Korea are concerned about energy security. They have to import a large percentage of the fossil fuels they use. Most solar and wind energy can be produced domestically and should support energy security objectives.

We will need to make investments in electrical transmission and distribution systems to link energy sources to consumers and accommodate the variability in the supply of renewable energy. We will need investments in energy storage to match supply and demand. A portion of annual energy industry investment is already spent on transmission and distribution.

Will the switch to renewables wreck the economy and cost jobs? We are talking about a transition that will take place over 30 years or more. There is time to make adjustments. There will be a natural trend with jobs moving over time from declining industries toward industries that are growing. There will be arguments over the details such as costs, technical problems, the effect on the economy and jobs, and other issues. The authors are convinced that we can solve or at least substantially reduce the impact of these problems if we want to. After reading this book, do you think we have a choice?

Chapter 16, The Way Forward, will discuss our recommendations as to what needs to be done.

CHAPTER 16

The way forward

It is time for action. Thirty years have passed since NASA scientist James Hansen testified before the U.S. Senate that global warming was a potentially serious problem and that it was primarily caused by the use of fossil fuels. Since that time not enough meaningful actions have taken place and greenhouse gas emissions have continued to rise. The facts presented in this book show the many dangers facing the earth due to global warming.

Further delays are unacceptable. So far, there has been a slow increase in the earth's temperature due to increases in greenhouse gas concentration in the atmosphere. At some point, as discussed earlier, we could reach a tipping point that will accelerate global warming in a potentially irreversible way, beyond our control. This is not a forecast. We do not know enough today to make a prediction. However, it is a real possibility we should not ignore.

There are not any quick fixes to global warming. There should not be a crash program setting unrealistic goals that cannot be met, attempting to solve this problem in a short period of time. As described in Chapter 14, Action Plan: Efficiency, Power, Transportation, and Land Use, we propose an *Action Plan* starting as soon as possible using existing technology to reduce global greenhouse gas emissions to about $47.2 \, GmtCO_{2eq}/year$ by 2050, essentially the level of global emissions in 2010. We know that this plan alone is not enough. Our suggestion is to start immediately with the technology we have today with the understanding that we need to find additional reductions as we proceed.

To *stop* global warming, it is essential to reduce greenhouse gas emissions to *net zero* as soon as possible. Additional reductions could include a larger carbon fee with greater annual increases, a more rapid transition to wind and solar, a more rapid transition to the electrification of transportation, an aggressive effort to plant more trees and restore forests, a more rapid upgrading of buildings, changes in agriculture, and a more rapid transition to energy-efficient equipment and appliances. More aggressive efforts to phase out coal, such as those proposed by Germany, would be beneficial.

Reaching Net Zero.
DOI: https://doi.org/10.1016/B978-0-12-823366-5.00016-6

© 2020 Elsevier Inc.
All rights reserved.

There is the potential for additional improvements associated with new technologies, such as carbon capture. However, it is questionable that any promising new technology can be demonstrated to be technically and economically feasible and implemented quickly enough on a large enough scale to make a big difference. We should assume for the immediate future that most of the needed improvements will come from more aggressive actions using the technology and approaches we have at hand.

There is the possibility that even more aggressive actions may become necessary if some tipping point is approached upon which global warming would accelerate beyond our ability to stop it. One such eventuality would be the release of large amounts of methane from melting permafrost.

We tend to be optimistic. The world wasted a lot of time in addressing global warming. If we start soon, we believe this problem can be solved even if the Intergovernmental Panel on Climate Change's (IPCC) goal to keep the global temperature rise under $1.5°C-2.0°C$ cannot be met. If the world continues to delay taking action, we are pessimistic about what world we will be leaving for our grandchildren.

The future can be bright

We can see a future where there are abundant sources of affordable energy in the form of electricity from renewables. Hydrogen and synthetic fuels are possible forms of green energy if they can be produced at reasonable cost using renewable energy. We believe it is possible to eliminate most greenhouse gas emissions and most air pollution and still have the energy needed to maintain and improve living standards and feed a larger global population. We do not need to make drastic lifestyle changes such as to have fewer children, to stop driving cars, or to reduce or eliminate air travel, or to all become vegetarians. Some activities will cost more, but there will be offsetting benefits as we have discussed.

If the United States and other governments make a strong enough commitment to stopping global warming and start taking the actions required, concerned and creative people and organizations will come up with practical solutions to the problems we face.

The transition to renewable energy and changes needed to combat global warming need not result in a financial crisis. There will be a major makeover of the energy sector of the global economy. However, this will take place over 30 or more years, giving time for the economy, individuals, and organizations to adjust. Most of today's energy infrastructure will become obsolete and will be replaced over the next 30–50 years anyway. In addition, much of our transportation system including almost all vehicles will be replaced by vehicles powered by electricity, hydrogen, or synthetic fuels. Buildings will be replaced or upgraded to be more energy-efficient. Most major appliances such as furnaces and hot water heaters will be replaced by heat pumps and other all-electric devices. All of this equipment will be designed to be more energy-efficient. There will be major new business opportunities for those who embrace the changes that will take place.

Top priorities

It is important that we do the right thing, not just do something. As we have stated earlier, if we do not do the big things, the small things do not count for much, even if they are useful and commendable. The small things by themselves cannot do enough to stop global warming. So, what are the big, must-do actions?

1. Educate the voting public so that they will support needed changes.
2. Institute a carbon emissions fee to offset fossil fuel subsidies, especially the cost of being allowed to discharge greenhouse gases and other pollutants into the atmosphere. A carbon fee should eliminate the need to subsidize renewable energy.
3. The United States must be part of the international effort to combat global warming. Global cooperation, especially between the United States, China, India, the European Union, and Japan, is essential to solve this global problem. The United States should become a leader in this effort. With the United States having the world's highest per capita energy use of the major countries, the United States has a responsibility to make the transition to renewable sources of energy and set an example for others.

In the following sections, we will discuss these top priorities, listing what we think needs to be done by federal and state governments, the public, and industry to defeat global warming.

Government actions

This list can also be used by concerned citizens as a checklist to see if their governments are doing what needs to be done.

- *Educating the public*

One of the highest priorities for solving global warming is the need for an informed public willing to support the changes needed. Politicians will express concern and even say that global warming is the most important issue. However, politicians will not take any actions that are not supported by their voting public.

For the United States, the first step in educating the public is for the federal government to acknowledge that global warming is real, is an immediate danger, and that something can and should be done about it. This is supported by scientific evidence and not an opinion. This acknowledgment would be similar to the Surgeon General's report issued in 1964, over 50 years ago, stating that the scientific data clearly linked cigarette smoking to lung cancer, heart disease, and other diseases. This report ended the debate over the health effects of smoking. It led to a number of actions to reduce cigarette smoking.

Similarly, we need the federal government to acknowledge that the scientific evidence clearly links greenhouse gas emissions to global warming and that global warming is leading to adverse climate changes and serious health and environmental problems that must be stopped. This recognition should go a long way toward ending further delays in taking action to reduce and eventually eliminate greenhouse gas emissions.

The scientific information supports this conclusion. In November 2018, the U.S. Global Climate Change Research Program issued *Climate Science Special Report*, stating that "It is extremely likely that human activities, especially emissions of greenhouse gases, are the dominant cause of the observed warming since the mid-20th century. There is no convincing alternative explanation supported by the extent of the observational evidence." This report was a joint effort of 13 federal government agencies including the Departments of Defense and Energy plus the Environmental Protection Agency, the National Aeronautics and Space Administration, and the Smithsonian Institution.[1]

In a separate announcement, the presidents of the National Academies of Science, Engineering, and Medicine stated that "Scientists have known for some time, from multiple lines of evidence, that humans are changing Earth's climate, primarily through greenhouse gas emissions. The evidence on the impacts of climate change is also clear and growing. The atmosphere and the Earth's oceans are warming, the magnitude and frequency of certain extreme events are increasing, and sea level is rising along our coasts."[2] In addition, they stated that, "A solid foundation of scientific evidence on climate change exists. It should be recognized, built upon, and most importantly, acted upon for the benefit of society."

We can see that the leading government organizations with knowledge of global warming conclude that global warming is real, is linked to fossil fuel use, and is a serious danger that must be addressed. What we need now is a clear statement from the federal government confirming the conclusions of these knowledgeable organizations.

- *Report card*

As part of educating the public, an easy-to-understand annual report card or the equivalent is needed, to be issued by an international agency or U.S. government organization. This would include a standard, easy-to-read format reporting on 10 or so key metrics that measure progress. Some examples could be annual global greenhouse gas emissions by type of gas, global temperature measurements, measures of greenhouse gas concentrations in the atmosphere and oceans, sea level rise, sea ice coverage, and a few others. There should be some measure of progress or lack of progress by the world's top sources of greenhouse gases. The growth of renewable energy should be included. Perhaps the top 10 most significant events related to global warming could be included.

---

[1] Wuebbles et al., *NCA4*.

[2] "National Academies Presidents Affirm the Scientific Evidence of Climate Change," *News, The National Academies of Sciences, Engineering, Medicine*, June 18, 2019, http://www8.nationalacademies.org/onpinews/newsitem.aspx?RecordID = 06182019.

This report card should be issued annually in a form that can be downloaded and printed by the general public.

- *Implement a fee on carbon emissions*

The biggest single thing that the United States could do to demonstrate that it is serious about doing something about global warming and also to set an example for others is to implement a national carbon fee. Charging a fee for carbon emissions is part of getting the incentives right. As the cost of carbon emissions is increased, renewable alternatives become more attractive to energy users without the need for subsidies.

As stated earlier, this fee should not be thought of as a tax but rather a fee for the discharge of CO_2 and other greenhouse gases into the atmosphere. We need a uniform national carbon fee, not a patchwork of 20 or more state fees. Congress should decide on the appropriate fee, any revenue sharing arrangements with the states, and how the money should be spent. There is already a bill before Congress to approve a carbon fee, HR 763, *Energy Innovation and Carbon Dividend Act of 2019.*

Part of carbon fee revenues should be allocated to low-income families, rather than as a payment to every citizen regardless of income. Payments, to the extent practical, should be to offset a family's actual increased spending associated with higher energy prices. This should not morph into another entitlement.

Most revenue from a carbon fee should be spent on important projects directly related to solving global warming. These should be projects such as major infrastructure projects that can only be implemented by the federal and state governments. This would be similar to the use of gasoline taxes to build and maintain the national highway system. If a large portion of carbon fees are not set aside for these government expenses, then projects would be financed by deficits that add to the national debt, posing an additional risk while attempting to deal with this global crisis.

- *Join the international effort to combat global warming*

The United States, China, India, the European Union, and Japan need to start discussing ways that they can lead the international effort to address climate change on a global basis. They need to cooperate and get other large economies to join in. This needs to be done despite disagreements in other areas. Without the United States and China working together, it is unlikely that the efforts of other countries will make enough difference. These two countries also have the largest technical resources needed to address global warming and are the ones who can most afford to make the investments required.

The United States cannot expect other countries to do more than what the United States is willing to do. Political parties have to make this a bipartisan issue and work together, no matter what their other interests and differences are. The United States must remain a member of the Paris Agreement and demonstrate leadership in dealing with global warming.

There will be a need to provide assistance to poor countries that are affected by global warming. For the most part, they are not responsible for the problems they have to deal with resulting from global warming. A portion of any carbon fee revenues from the countries that are the largest users of fossil fuels should be used to assist developing countries that do not have the needed technology or finances to stop global warming or mitigate its effects.

Most deforestation and much of the emissions from farming are from developing countries. These countries will need technical and financial aid to help them modify farming and land-use practices that are major contributors to greenhouse gas emissions. The destruction of carbon sinks such as tropical forests needs to be stopped. This could include financial incentives such as payments to protect and expand forests.

- *National plan*

  Every country should develop a plan. For the United States, Congress, the executive branch, and federal agencies should work together to develop a national plan to deal with global warming that includes a roadmap and timeline to get to net zero. The states should be asked to contribute to this plan. This plan should be realistic and practical, not idealistic. It should not set goals that cannot be met. It should estimate funding requirements to reduce greenhouse gas emissions to net zero. Each state should be asked to develop a plan consistent with the federal plan.

  Committees should be established in the House and Senate dedicated to the global warming problem. This would be better than assigning global warming to existing committees that already have other responsibilities that would limit their ability to focus on this threat. Public hearings to explore ways to reduce greenhouse gas emissions should be held. These hearings could be an important part of educating the public.

To implement any project successfully, there needs to be someone and some organization with the clear authority needed to do the job. We do not need another cabinet-level department to address global warming. This task could be the responsibility of the Department of Energy, which should have the technical resources needed at the federal level. Several other government organizations should also be involved, such as the Environmental Protection Agency. We should determine how best to coordinate our overall federal government response. Perhaps changes implemented after 9/11 to coordinate the government's response to international terrorism would have some useful features worth duplicating. The responsibilities of all government organizations involved should be clearly stated along with clear lines of communication and well-defined authority needed for implementation. Any plan should be updated regularly, say every 2 years, as understanding what it takes to get to net zero improves.

- *Technology plan*

  Every country needs a roadmap or plan to identify the capabilities necessary to maintain technologies that are key to the country's future economic growth and national defense. For example, as manufacturing jobs and facilities are relocated abroad, the United States becomes increasingly reliant upon foreign sources for technology, materials, and components. China is a rival but not necessarily an adversary. China is the first country that has the potential to match or exceed U.S. capabilities in technology and gross domestic product. China has a potentially larger domestic market because of its huge population and economic growth. This could give China greater economies of scale such as in electric vehicle production and also attract foreign investment and greater technology transfers as a price to access this market.

  China has a national plan, "Made in China 2025," that was launched in 2015 with the purpose of modernizing China's economy to close technology gaps, to become the world leader in key technologies, and to lessen China's dependence upon foreign technology. The plan is also to position China to become the world leader in certain industries. Although we can be skeptical as to the success of such a plan, we should not assume that China will fail. China is already leading the world in electric vehicle production, nuclear power, and the production and installation of solar and wind power.

As the U.S. economy transitions to renewable energy, it would be unfortunate if the United States and European Union had to import batteries, electric vehicles, solar arrays, and other products from China and miss many of the business opportunities associated with renewable energy.

- *Siting*

One of the biggest issues that led to the failure of nuclear power was plant siting. For renewables, about 25% of solar arrays can be installed on rooftops and are unlikely to be a big problem. However, there should be a national policy or guidelines for siting large solar and wind farms. Solar and wind do not have the same safety issues as a nuclear power plant or facilities to process and store nuclear waste. However, large solar and wind farms with their associated transmission lines are unsightly to many and would not be welcome near populated or scenic areas.

Any plan should identify those areas that are the best locations for large wind and solar farms. They should be located where they would be most effective with transmission lines transporting electricity from where it is most efficiently produced to where it is needed. For example, Texas and Arizona are obviously better locations for large solar farms than New England. (Refer to discussion of Texas and the Electric Reliability Council of Texas in Chapter 13: Some Success and Failures.) This is similar to what we do today with fossil fuels that are usually found in remote locations and transported to where they are needed. Transmission lines for renewable electricity are the equivalent of today's pipelines, coal trains, and oil tankers.

Public lands could be leased for major solar and wind farms as the government leases federal lands today for oil and gas exploration and development. Some military bases and various test sites have large buffer zones that could be used for solar or wind farms without compromising their military mission.

- *New laws and regulations*

New laws and regulations will be needed to address global warming. Regulations are needed so that everyone clearly knows what needs to be done and what is allowed or prohibited. Congress and the states should be sure that legislation and regulations are clearly written and easy to understand.

- *Demonstration projects, subsidies, mitigation, and other actions*

States or even other countries could serve as large-scale demonstration projects to determine the best way to proceed. State efforts should be monitored and their successes should be emulated. China is rapidly expanding its fleet of electric vehicles from scooters to automobiles, trucks, and buses. They will pioneer ways to accommodate a large fleet of electric vehicles in an urban environment. In the United States, the federal government should support renewable energy projects in Hawaii, California, Texas, and other states as large-scale demonstration projects. California and Hawaii have a goal to produce all electricity used in the state from renewables by 2045. In 2018, governors in Connecticut, Illinois, and Nevada voiced a commitment to move to 100% renewables. In 2019, legislatures in New York and Maine approved bills to have electricity from 100% renewables by 2040 and 2050, respectively. Internationally, Albania, Iceland, Paraguay, and Norway are at 100% renewable energy or nearly there. If successful, these examples could be national and international achievements to emulate.

A good public investment would be to support some large-scale demonstration projects and fund some state efforts that would work out technical details and determine the economics of various solutions. Perhaps there could be a federal program modeled after the Defense Department's DARPA organization (Defense Advanced Research Projects Agency) established to fund research and development for national defense.

Some large-scale infrastructure projects should be funded by the federal government. Some possibilities are:

- national and regional plans for a transmission network similar to the national highway system. This could be a national or regional version of what Texas did to connect west Texas to cities in the east;
- projects to recycle CO_2 or to remove CO_2 from the atmosphere and convert the CO_2 into useable fuels using electricity from renewables;
- smart electrical grids to match electricity demand with variable supplies from renewables;
- large-scale electricity storage systems to help offset the variability of renewables.
- *Subsidies should be eliminated*

This is an important step. It should include any subsidies or mandates for ethanol. Incentives in the form of mandates, subsidies, or tax breaks should be minimized. They distort the market. Subsidies and other government benefits often become permanent entitlements for special interests. The major change is the need for a carbon fee to offset present subsidies for fossil fuels and level the playing field with renewables. Ethanol has benefited from subsidies and mandates for years and should be able to survive without them if it is a cost-effective fuel. It is time to let the market decide.

As reported in this book, the cost of renewable energy is declining rapidly and is increasingly cheaper than fossil fuels or nuclear energy. Yes, there are problems to solve, such as accommodating the variability of electricity from wind and solar. However, the cost of renewable energy is declining steadily due to competition among suppliers, increasing economies of scale, the evolution of solar arrays and wind turbines, and new technologies. Governments should be sure that the declining cost of renewables is passed on to customers. Regulations that increase the cost of renewable energy, such as a requirement to pay an above-market price for electricity from renewables are counter-productive.

We recommend a policy to *let failures die*. Some projects will earn the right to die. We have presented several examples of well-meaning but poorly structured programs that ended in failure. In addressing global warming, some mistakes will be made—that is unavoidable. The problem occurs when we are faced with a failure or marginal success and do not do anything about it. It is important to cut our losses to free up resources for programs that are succeeding. The worst outcome is to subsidize failures by maintaining mandates and subsidies that only prop up or reward politically connected special interests. Ethanol is a good example.

Among practical programs that might be considered is the equivalent of a "cash for clunkers" program to get inefficient vehicles off the road and to replace appliances such as refrigerators, dish and clothes washers and air conditioners with more efficient units. Measures should be supported to replace gas-fired furnaces and hot water heaters with heat pumps and electric hot water heaters using electricity from renewables.

As mentioned previously, it is important that the declining cost of electricity from renewable sources is passed on to consumers. The cost differential in favor of renewables should be strengthened by returning any savings due to their declining costs. Texas is a good example of what can be done by deregulating the electricity market so that there is price competition between suppliers. Texas's electricity prices are below the national average and far below electricity rates in California where rates are set by a public utility commission.

- *Mitigation*

Mitigation is another element of government programs that needs consideration. Offsetting the effects of global warming will become a big problem even if aggressive steps are taken to stop it. There are a number of actions that need to be taken at the state and national levels for mitigation. This could include better planning for disaster relief and infrastructure projects such as seawalls and levees to reduce flooding. However, it should be emphasized that mitigation has near-term benefits and is not a substitute for actions that reduce greenhouse gas emissions. As part of mitigation, we will have to face the fact that it will not be economical to preserve a number of inhabited locations by holding back flooding and rising sea levels. We need to be willing to abandon some locations where the cost of preservation is much higher than the cost of relocation. The government needs to decide what to save and what to abandon and at what cost. For example, who will bear the cost of abandoning or relocating property in coastal areas and flood plains that are threatened by a rising sea level or more severe storms? What happens when these properties cannot be sold or have to be sold at substantial discounts? There are cost trade-offs to be made. We should not spend unlimited funds trying to protect beach front property that should be abandoned.

Actions for concerned citizens

- *Educate yourself and others*

Educating the public was one of the major reasons for writing this book. Contact your government representatives, and most importantly, vote for candidates who are serious about addressing global warming. Spread the word. If this book is helpful, let others know about it. It is available from your favorite bookseller. Educate yourself. Some suggestions for further reading are found in the Appendices.

- *Engage with climate change activists*

  As we stated elsewhere in this book, significant progress will not be made in reducing greenhouse gas emissions until a better-informed public demands action by government and industry leaders. There are advocacy groups that interested individuals can join or support to help bring about this change. These include international student groups and others that are demanding action on a global basis. Appendix 7 has descriptions of some of the principal groups. One such group in the United States is the Citizens' Climate Lobby (CCL). CCL is a nonprofit, nonpartisan, grassroots advocacy organization focused on national policies to address climate change (see Appendix 6 for details).

- *Lobby for government action*

  Lobby for the United States to actively participate as a member of the Paris Agreement. The United States needs to be part of the international effort to deal with global warming and its effects. Actively lobby state and federal representatives to be sure they are taking useful actions to stop global warming. Be sure to ask candidates for government positions to clearly state their position on global warming and what they plan to do about it. The biggest single program to support is one that imposes a carbon fee and eliminates fossil fuel subsidies.

- *What about litigation?*

  As citizens, should we support litigation? It is not something we recommend. Lawsuits will not stop global warming, but they may help by calling attention to its serious health risks and the failure of governments to take appropriate actions to mitigate risks. Commencing in 2015, a virtual swarm of litigation has targeted major oil companies and the U.S. government. Plaintiffs range from a group of children who are suing the federal government under the "public trust doctrine," a legal concept that says that the government holds resources such as water, land, and air in trust for its citizens, to city, county, and state governments that are suing fossil fuel companies to recover public health and property damage costs caused by global warming. We provide a short summary of major lawsuits in Appendix 6.

- *Divest from major carbon emitters*

  Beginning in 2012, the environmental group 350.org, founded by Bill McKibben, began urging investment funds and others to divest investments in the major greenhouse gas emitters such as Shell Oil,

ExxonMobil, and others. To date, more than 1000 pledges on behalf of investment funds worth more than \$8 trillion have been recorded. In 2019, two-thirds of the members of the University of California Academic Senate voted to divest the University's \$12 billion endowment of all major fossil fuel companies.[3] The University took this step in May, 2020. While divestment might seem like a symbolic action, it could actually reduce investment risk. As renewables replace fossil fuels, the fossil fuel companies may be left with stranded assets. This could greatly reduce their stock prices. Litigation and claims are another risk. Chubb, the world's largest publicly traded insurance company, with operations in 54 countries, announced that it will no longer provide insurance to new coal-fired power plants or sell new policies to companies that derive more than 30% of their revenues from thermal coal mining.[4]

- *Be aware of the changing political landscape*

In many countries, politicians running for election are endorsing actions regarding climate change as a goal. Nowhere is this more evident than in the United States, the world's second largest emitter of greenhouse gases. The midterm elections of 2018 saw dozens of new, mostly young members, many female, elected to the U.S. House of Representatives. A flurry of legislative action, some of it related to climate change, began immediately upon their taking office.

On the political front, one of the most important actions that can be taken by the public is to learn what political candidates' ideas are on how to combat global warming and to vote for those who will take meaningful action.

Finally, we have seen flurries of excitement in the past that faded away. Let us hope that the events described above portend a sea change in political thinking and commitment. In an article published concerning the political situation in the spring of 2019, Bill McKibben wrote, "I think I can say that we are in a remarkable moment, when after years of languishing, climate concern is suddenly and explosively rising to the top of the political agenda."[5]

---

[3] Joshua E. Smith, "UC Regents are Urged to Divest from Fossil Fuels," *Los Angeles Times*, July 18, 2019, p. B3.

[4] Oliver Ralph, "U.S. Insurer Chubb Pulls Back from Coal," *Financial Times*, June 30, 2019.

[5] Bill McKibben, "Notes from a Remarkable Political Moment for Climate Change," *The New Yorker*, May 1, 2019.

- *Other individual actions*

Be willing to personally accept cost increases for gasoline and other fuels and for restrictions that may be required to address global warming. A higher price for fossil fuel use is probably the most obvious sacrifice we will all need to make. Increases will be temporary until fossil fuels are replaced by renewable energy.

Pay attention to how government funds are used. We cannot afford to waste resources on marginal or failing efforts or subsidies. Watch the money. Be sure that any funds raised by a carbon fee are well spent! Do not let revenues from a carbon fee be part of the general fund to be spent by politicians on purposes not clearly related to global warming.

Make lifestyle changes; "walk the walk." Pay attention to your energy use and carbon footprint and take appropriate actions, such as retrofitting your home with solar energy, installing LED lighting, buying an electric vehicle, etc. There are many practical actions that individuals can take.

Actions for industry

- *Keep innovating*

Industry is both an innovator of energy-efficient devices and technology and a major energy user. Industry led the way after the 1973 oil embargo by developing and embracing new energy-efficient technologies. Anyone wanting proof of this only has to recall what it was like to buy a low-wattage fluorescent lamp in 1975. Your local hardware store might have had two or three types. Today, at a large home improvement store, one encounters an aisle 100 yards long with thousands of LED lamps, smart thermostats, occupancy sensors, and dozens of other energy-saving devices. There are energy-efficient refrigerators, air conditioners, and other appliances. More such products will be available in the future.

As a major energy user, industry needs to continue to adopt energy-saving methods and processes in manufacturing. Experience shows that conservation and efficiency improvements are always possible and are good investments. Some of this will no doubt be driven by rising fossil fuel costs. Reducing the energy component of manufacturing costs will make or keep products competitive. Improving efficiency and conservation also applies to service industries that are now the largest share of the U.S. economy and major energy users.

- *Changing corporate philosophies*

  Industry should not fight the switch toward greater use of electricity from renewable sources. Invest in the future, do not try to preserve the past. Come up with a plan for your business to fight global warming. This will lead to good public and employee relations as the world becomes convinced that global warming is real and that something can be done about it.

  Our large energy companies and their suppliers and subcontractors should be leaders in the switch to renewable energy. We need the expertise of today's large energy companies, the engineering and construction firms serving the industry, and large electric utilities. They could play a major role in reducing global warming. The world's energy industry needs to be transformed. The big oil companies and other companies involved with the production and use of fossil fuels need to become major renewable energy suppliers. These companies have the financial and technical resources and experience to plan, build, and operate large energy projects anywhere in the world. If they can drill for oil in the North Sea, Alaska, or the Middle East, they should be able to plan and implement mega projects that will be needed, such as building solar farms in the Sahara and transporting electricity to Europe. They need to become major producers of hydrogen and synthetic fuels made from captured CO_2 using renewable energy to power the process.

  NextEra Energy Resources could be a good example of what is needed. NextEra is now the world's largest generator of renewable energy from wind and solar. They build wind and solar farms, electric transmission lines, and energy storage facilities. NextEra owns Florida Power and Light, a traditional electric utility, which is the largest regulated electric utility in the United States. This is a good example for others to follow in embracing the changes required to reach net zero.

- *Many companies are proactive about fighting global warming*

  Many companies balk at the idea of governments ignoring global warming. Instead, they take the high road by taking actions that they believe will appeal to the majority of their customers. There is evidence that "doing good" can actually lead to more profitable operations. The most important areas, identified by many major corporations, include better treatment of employees, customers, suppliers, improved product quality, response to environmental concerns, community participation, and job creation. In a recent publication, Forbes magazine listed 100 of America's best corporate citizens.[6] Among the top 20 companies in the environmental awareness category one finds Microsoft, IBM, Xerox, Cummins, and Bank of New York Mellon. All in all, 890 companies were analyzed, and the article also reports the bottom 10%.

The idea of changing corporate responsibility is also reflected in corporate annual reports. Today, many major corporations provide annual "Sustainability Reports" to shareholders to keep them informed of the corporation's goals to improve energy efficiency and reduce greenhouse gas emissions. We conducted a brief review of 20 annual reports from 2016 to 2017. Some of the measures these companies reported implementing included energy and efficiency improvements with specific goals for reducing energy use in facilities and operations, purchasing renewable energy, setting greenhouse gas emission reduction goals, investing in wind power projects, and improving vehicle fleet fuel efficiency. Sample comments from these reports are reproduced in Appendix 8.

Another very positive move by major global corporations is joining RE100 (an acronym for Renewable Energy 100%). Companies that affiliate with RE100 establish a public goal to meet 100% of their corporation-wide electricity demand from renewable sources within a specified timeframe, by 2050 at the latest. They also commit to disclose electrical use annually so their progress in meeting their goal can be monitored.[7] RE100 began at Climate Week NYC 2014. Since then, the initiative has grown to include major companies in Europe and North America, India, China, Japan, and Australia. Membership encompasses a wide range of industries from telecommunications and information technology to cement and automobile manufacturing.

- *Corporate naysayers*

    Unfortunately, this concern for reducing global warming is not reflected in the activities of many major global organizations and corporations. A common refrain is that "we are all responsible for global

---

[6] Maggie McGrath, "Meet America's New Regulator: Adam Smith," *Forbes*, December 31, 2018, pp. 55—68.

[7] "RE100, "The World's Most Influential Companies, Committed to 100% Renewable Power," http://there100.org.

warming," or "we are just providing a product that is demanded by the public." Saying that "we are all responsible," is a specious argument in that we cannot really compare the consequences of a Chevron Corporation's business with a small hardware store in middle-America or a European executive who flies to Asia on business. We are beginning to see awareness of the need to address the irresponsible actions of large corporations that materially affect global warming. Public sentiment seems to be shifting in the direction of demanding responsibility for environmental pollutions, more frequent storm damage, and wildfires attributed to climate change. One author writes, "If even one judge finds a fossil fuel company liable for climate-related damages, that will open the floodgates."[8] This resistance is taking many forms. Legal measures are costly and large corporations have the resources to fight back. On the other hand there is a growing movement of shareholder protests that include unloading stocks from the worst offending companies. Banks are becoming increasingly nervous about loans for projects that may be subject to declining revenues or increased regulatory scrutiny. No company wants to find itself listed in a national publication as being part of the "worst of the worst."

- *There is a small group of bad actors*

Let us start with the bad news. Just 100 global companies have been the source of more than 70% of the world's greenhouse gas emissions since 1988—the year the IPCC was formed.[9] In other words, a small group of international companies could have a major impact on carbon emissions—if they chose to. Instead, they continue to invest in exploration and resource development projects that may ultimately become worthless, as renewable energy has no exploration ("dry hole") risks and becomes cheaper to produce.

As criticism has mounted, the oil companies publicize that they are determined to be "part of the solution." They are investing in carbon dioxide capture and storage projects that sound good in television ads but do not eliminate combustion.

Jason Mark's "The Climate Wrecking Industry" article referenced above lists the "Worst of the Worst": Koch Industries, ExxonMobil, 21st Century Fox, Berkshire Hathaway, Chevron, Shell, Duke Energy,

---

[8] Ann Carlson, UCLA Law Professor, quoted in Jason Mark, "The climate wrecking industry: and how to beat it," *The Nation*, September 24/October 1, 2018, pp. 14–17.

[9] Tess Riley, "Just 100 Companies Responsible for 71% of Global Emissions, Study Says," *The Guardian*, February 2018.

Southern Company, British Petroleum, American Electric Power, and General Electric, to name a few.

No surprises here. Just read the annual report excerpts in Appendix 8 for Duke Power, Chevron, and the Southern Company.

What next?

Hopefully, the reader will find this book useful and alarming. The authors believe that the technical and financial resources are or could be available to solve this problem. What we need is public support for action with cooperation and coordination among the largest countries contributing to global warming today.

Do we have a choice? Failure is an option. For many practical reasons, the United States or other countries that are major sources of greenhouse gas emissions may not do enough to reduce emissions to net zero. The problem is not going away. It only gets bigger and harder to solve. If we delay, eventually the obvious effects of global warming will be impossible to ignore. If we do not take the initiative to address this threat, it is possible that some parts of the world will experience a global warming disaster that will be an unavoidable wake-up call for the United States and other countries. The timing and nature of such an event are impossible to predict.

Success is possible. We are optimistic and hope that global warming will be addressed in time to prevent more damage and before the cost of mitigation becomes impossibly high.

Afterword

Global warming and associated climate change is a severe and unique challenge to the United States and to the world. The problem is growing and will only get worse with time. We can take action now or wait until more irreversible damage has been done and more aggressive actions are required. A delay also increases the chance of reaching a tipping point after which global warming could accelerate or we have some climate catastrophe that forces the major world economies to act.

We authors were born just before the United States entered World War II. The Great Depression of the 1930s and that war had a big influence on our parents' and grandparents' views of the world. It was a lesson about what our country could accomplish when it was fully united to solve a problem or to meet a threat to the nation and the world.

We can remember the end of World War II, the atomic bombs dropped on Japan, and the atmospheric testing of hydrogen bombs by the United States and Russia.

In junior high school we could watch the latest news about the Korean War on an amazing new invention, television.

During our first year as engineering students, *Sputnik* was launched by Russia and started the space race. The United States won, placing a man on the moon in 1969. We succeeded but that victory was not a foregone conclusion.

We completed our B.S. degrees using slide rules, not computers.

The Vietnam War with its domestic turmoil and political strife is a vivid memory. One of us served in the Navy during this war and was stationed in Washington, D.C. during the Cuban Missile Crisis and the Kennedy assassination. One of us was involved in the renovation of the Pentagon, before and after 9/11.

We both spent part of our careers in the promising nuclear power industry and personally saw the rapid rise and eventual failure of commercial nuclear power. Failing big is possible even with the best intentions. So far, we have seen the invention and widespread use of jet planes, television, a polio vaccine, nuclear weapons, the national highway system, space travel, intercontinental ballistic missiles, medical imaging, the computer, digital communications, the Hubble telescope, the Internet, genomics, cell phones, and many other amazing and sometimes frightening things. We were surprised and pleased that the Cold War ended without a military conflict. We have seen great progress as well as threats to our lives and the environment.

In spite of wars, disease, population growth, recessions, and other happenings there has been progress. We cannot rule out failure or even a catastrophe. However, we are optimists betting on eventual success with some big bumps along the way.

Ironically, as we finished editing this book in mid-2020, the world was reeling from the Covid-19 pandemic, with its disastrous effects on health and the global economy. The pandemic is a wakeup call to what global warming could do. With the pandemic, we will eventually have a vaccine and the economy will recover. With global warming there is no vaccine—no solution other than reaching net zero.

We are trying to present practical actions that must be taken to reduce global warming, knowing that the problem may be much more serious and more difficult to solve than we are representing in this book. Even if what we recommend is not enough, it is an important step in the right direction, and the effort and expense will not be wasted.

We have a concern that the United States has lost its pioneering spirit and the will to imagine transformative projects: the transcontinental railroad, the Panama Canal, the Manhattan Project, and the Apollo program. A number of new laws and regulations will be needed. However, global warming cannot be solved by long and complicated legislation with a lot of mandates, subsidies, and exceptions for special interests. Legislation and regulations should be kept simple and easily understandable by the public, the government agencies responsible for implementation, and those affected by them. Litigation may not help and could even be counterproductive, except for drawing public attention to polluters and establishing the need for tougher regulations. We should focus on the future, not past actions unless there was criminal intent.

To succeed, we need dreamers, builders, and innovators, not bureaucrats. We need to be united in our efforts to tackle this complex problem, not be divided and distracted by those who disbelieve and discredit science. Most important, we need to be part of a coordinated global effort to solve this difficult and threatening problem.

Bill Fletcher and Craig Smith

Further reading

Doe, Robert (2006). *Extreme Floods: A History in a Changing Climate*. Gloucestershire, UK: Sutton Publishing. Doe points out that floods kill more people and damage more property than any other natural phenomenon. European Environment Agency data show that two-thirds of all catastrophic events since 1980 were directly attributable to weather and climate extremes. Humans need to learn to manage the risks.

Flannery, Tim (2005). *The Weather Makers: How Man Is Changing the Climate and What It Means for Life on Earth*. New York, NY: Grove Press. This international best seller lays out the overwhelming evidence that humans are impacting climate in a way that portends disaster unless immediate actions are taken to reduce greenhouse gas emissions. Also describes the lack of political willpower to address the problem on a global scale.

Goldsmith, C.S. (2007). *Uninhabitable: A Case for Caution*. Las Vegas, NV: Goldstar publications. Author Goldsmith says he is not a scientist, but a 1998 CIA report on global warming caught his attention and launched a 5-year research program. He has studied the available information carefully and has written an excellent book, including a frightening, but plausible future forecast.

Goodell, Jeff (2017). *The Water Will Come: Rising Seas, Sinking Cities, and the Remaking of the Civilized World*. New York, NY: Little, Brown and Company (Hachette Book Group). In this sobering book, Goodell describes how, around the world, rising seas are pushing water into the places where we live. By century's end, tens of millions of people will be retreating from coastal areas.

Hansen, James (2009). *Storms of My Grandchildren: The Truth About the Coming Climate Catastrophe and Our Last Chance to Save Humanity*. New York, NY: Bloomsbury USA. In 2001, Dr. James Hansen, NASA scientist, warned a high-level task force of the Bush administration that climate change was a real and present danger. He was ignored, and the nation has now lost two decades in which corrective measures could have been undertaken.

Hawken, Paul (ed.) (2017). *Drawdown: The Most Comprehensive Plan Ever Proposed to Reverse Global Warming*. New York, NY: Penguin Books. Hawken assembled a team of international physical and social scientists and economists to evaluate 80 measures for reducing greenhouse gas emissions in seven broad categories (buildings, transportation, agriculture, etc.) and to estimate costs and savings. While some measures are based on proven technologies, others involve long-term cultural changes and are unlikely to have significant impact.

Iyengar, Kuppaswamy (2015). *Sustainable Architectural Design: An Overview*. New York, NY: Routledge. In this classic work, Professor Iyengar describes sustainable building design using renewable energy sources that minimize fossil fuel use and greenhouse gas emissions.

Jones, Lucy (2018). *The Big Ones: How Natural Disasters Have Shaped Us (and What We Can Do About Them)*, New York, NY: Doubleday. Noted Caltech scientist Lucy Jones reviews historic natural disasters (fires, floods, earthquakes, and volcanos) and suggests measures to minimize economic costs and loss of life.

Kress, John W., et al. (2017). *Living in the Anthropocene: Earth in the Age of Humans*. Washington, DC: Smithsonian Institution Press. This book postulates that the earth is in a new era—that of humans—that is fundamentally changing it. Thirty essays by scientists, historians, archaeologists, and anthropologists describe problems of population growth, scarcity of resources, climate change, and environmental pollution and speculate on what it means for our future.

Masri, Shahir (2018). *Beyond Debate: Answers to 50 Misconceptions on Climate Change*. Newport Beach, CA: Dockside Sailing Press. In clear and nonscientific language, Dr. Masri explains the facts behind many of the most common misconceptions regarding climate change.

Marshall, George (2014). *Don't Even Think About It: Why Our Brains Are Wired to Ignore Climate Change*. New York, NY: Bloomsbury USA. Addresses why, in spite of overwhelming scientific evidence, the threats from climate change have been largely ignored to date.

McKibben, Bill (2010). *Eaarth: Making a Life on a Tough New Planet*. New York, NY: Times Books. McKibben writes eloquently how human failure to take action on global warming is creating a new world characterized by rising seas, spreading of disease, and other dangers. He urges that we adopt a more sustainable lifestyle that will enable us to weather climate change while transitioning to a better future.

McKibben, Bill (2019). *Falter*. New York, NY: Henry Holt and Company. Discusses the irreversible changes that are damaging the earth and actions needed to save the world from climate change.

Smith, Craig B. and Parmenter, Kelly E. (2016). *Energy Management Principles: Applications, Benefits, Savings* (2nd ed.). Oxford, UK: Elsevier. Drawing upon 35 years of practical experience since the first edition was published, Smith and Parmenter describe hundreds of ways to use energy more efficiently and reduce greenhouse gas emissions.

Wallace-Wells, David (2019). *The Uninhabitable Earth: Life After Warming*. New York: Tim Duggan Books. A discussion of life on earth if global warming is not addressed. Discusses the impact on life in the future if this problem is not solved.

Useful reports

1. Fossil CO_2 and GHG emissions for all the world countries, JRC Science for Policy Report — edgar.jrc.ec.europa.eu/overview.php?v=booklet2019

2. International Energy Agency (IEA): Key World Energy Statistics — www.iea.org/reports/world-energy-statistics-2019

3. International Energy Agency (IEA) World Energy Outlook 2019 — www.iea.org/reports/world-energy-outlook-2019

4. CO_2 and Greenhouse Gas Emissions, Our World Data — ourworldindata.org/co2-and-other-greenhouse-gas-emissions

5. Global Energy Transformation: A Roadmap to 2050, International Renewable Energy Agency (IRENA) — www.irena.org/publications/2019/Apr/Global-energy-transformation-A-roadmap-to-2050-2019Edition

6. Global Energy System based on 100% Renewable Energy (2019), Energy Watch Group — energywatchgroup.org/new-study-global-energy-system-based-100-renewable-energy

7. Lazard's Levelized Cost of Energy Analysis—Version 13.0 — www.lazard.com/media/451086/lazards-levelized-cost-of-energy-version-130-vf.pdf

8. Lazard's Levelized Cost of Storage Analysis—Version 5.0 — www.lazard.com/media/451087/lazards-levelized-cost-of-storage-version-50-vf.pdf

9. Energy and Climate Change: Challenges and Opportunities — https://www.youtube.com/watch?v=w4vtJWKF3E8

10. U.S. Climate Science Special Report, Fourth National Climate Assessment (NCA4),15 Global Change Research Program (USGCRP) — www.ncei.noaa.gov/news/usgcrp-climate-science-special-report

11. The Climate Report. U.S. Global Change Research Program — www.mhpbooks.com/books/the-climate-report/

12. Climate Change Is the Greatest Threat to Human Health in History, Health Affairs Blog — www.healthaffairs.org/do/10.1377/hblog20181218.278288/full/

13. Climate Change in a Nutshell: The Gathering Storm — www.columbia.edu/~jeh1/mailings/2018/20181206_Nutshell.pdf

14. BP Energy Outlook 2019 edition — www.bp.com/content/dam/bp/business-sites/en/global/corporate/pdfs/energy-economics/energy-outlook/bp-energy-outlook-2019.pdf

15. ExxonMobil Outlook for Energy: A Perspective to 2040 — corporate.exxonmobil.com/-/media/Global/Files/outlook-for-energy/2019-Outlook-for-Energy_v4.pdf

The Intergovernmental Panel on Climate Change (IPCC) Special Reports:

16. Special Report on Global Warming of 1.5°C (SR1.5) — www.ipcc.ch/sr15/
17. Special Report on the Ocean and Cryosphere in a Changing Climate (2019) — www.ipcc.ch/srocc/
18. Special Report on Climate Change and Land (2019) — www.ipcc.ch/srccl/

McKinsey & Company Reports:

19. Global Energy Perspective 2019: Reference Case. Energy Insights by McKinsey (January 2019) — www.mckinsey.com/industries/oil-and-gas/our-insights/global-energy-perspective-2019
20. Global Energy Perspective: Accelerated Transition. Energy Insights by McKinsey (November 2018) — www.mckinsey.com/industries/oil-and-gas/how-we-help-clients/energy-insights/global-energy-perspective-accelerated-transition
21. Game Changers in the Energy System: Emerging Themes Reshaping the Energy Landscape. World Economic Forum in collaboration with McKinsey & Company (January 2017) — www3.weforum.org/docs/WEF_Game_Changers_in_the_Energy_System.pdf

Useful websites

Bloomberg Carbon Clock	www.bloomberg.com/graphics/carbon-clock/
Climate Central	www.climatecentral.org/
Carbon Brief	www.carbonbrief.org/
Carbon Tracker Initiative	www.carbontracker.org/
Global Carbon Project	www.globalcarbonproject.org/
Our World Data	ourworldindata.org/
Emissions Database for Global Atmospheric Research (EDGAR)	edgar.jrc.ec.europa.eu/
International Renewable Energy Agency (IRENA)	www.irena.org/
Intergovernmental Panel on Climate Change (IPCC)	https://www.ipcc.ch/
U.S. National Oceanic and Atmospheric Administration (NOAA)	www.noaa.gov/
U.S. Environmental Protection Agency (EPA)	www.epa.gov/
Copernicus Atmospheric Monitoring Service	atmosphere.copernicus.eu/
Global Emissions Data, Center for Climate and Energy Solutions	www.c2es.org/content/international-emissions/
Enerdata Global Energy Statistical Yearbook	www.enerdata.net/
Center for Climate and Energy Solutions, Global Carbon Atlas	www.globalcarbonatlas.org/en/content/welcome-carbon-atlas

PART IV

Appendices

APPENDIX 1

Abbreviations, units, and conversion factors

Abbreviations

bbl	(unit)	Oil barrel
Btu	(unit)	British thermal unit
CH_4	(abbreviation)	Methane gas
CO_2	(abbreviation)	Carbon dioxide gas
CO_{2eq}	(abbreviation)	Carbon dioxide gas equivalent (includes $CO_2 + CH_4$ and other GHG)
ft.	(abbreviation)	Foot
G	1,000,000,000	Giga (10^9)
GDP	(abbreviation)	Gross domestic product
GHG	(abbreviation)	Greenhouse gas(s)
Gmt	(abbreviation)	Gigametric ton
g	(unit)	Gram
gal	(abbreviation)	Gallon
h	(unit)	Hour
ha	Hectare	$10^4 \, m^2$
hp	(abbreviation)	Horsepower
i	interest rate	Dimensionless
J	(unit)	Joule
k	one thousand	Kilo (10^3)
kg	(abbreviation)	Kilogram
km	(abbreviation)	Kilometer
lb	(unit)	Pound
M	1,000,000	Mega (10^6)
MBtu	(unit)	10^6 Btu
m	(abbreviation)	Meter
mpg	(abbreviation)	Miles per gallon
mph	(abbreviation)	Mile per hour

(*Continued*)

(Continued)

mt	(abbreviation)	Metric ton
NOx	(abbreviation)	Oxides of nitrogen
st	(abbreviation)	Short ton (2000 lbs)
therm	(unit)	10^5 Btu
toe	(abbreviation)	Metric ton of oil equivalent
quad	(unit)	10^{15} Btu

Units and conversion factors

This book has been prepared using the international system of units (SI). In SI practice, the approved units for energy and power are the Joule and the Watt:

Energy, heat, and work: Joule (J) = 1 Newton-meter = 1 Watt-second.

Power: Watt (W) = 1 Joule/second = 1 Newton-meter per second.

When SI units are used for calculations, they are frequently preceded by a prefix that is a multiple of 10. Then the same basic unit can be used to measure a very large or a very small quantity. Thus one meter is slightly more than a yard, while a kilometer is slightly more than 1/2 mile. The prefixes, symbols, and multipliers encountered in this book are:

kilo (k), 10^3 (thousand) = 1000

Mega (M), 10^6 (million) = 1,000,000

Giga (G), 10^9 (billion) = 1,000,000,000

Tera (T), 10^{12} (trillion) = 1,000,000,000,000

These large units can be hard to visualize. For example, the United States currently emits 1.0 Gmt (gigametric tons) of CO_{2eq} into the atmosphere every 2 months. That is a billion metric tons, or 2,200,000,000,000 pounds. Consider instead that this is equal to the weight of 150 *million* male African elephants, and you get a better feeling for the enormity of the problem. You know that CO_2 is a gas and therefore is lightweight. It takes a lot of CO_2 to weigh one metric ton. In fact, it is an amount equivalent to that which theoretically could be contained in about 7.5 of those large 40 ft. containers you see on container ships! And that is just for a single metric ton

The SI system of units has several important advantages: (1) it is universal, with a single unit for each quantity (e.g., m for length); (2) it uses decimal arithmetic, facilitating changes, calculations, and conversions; and (3) it is coherent, meaning that when two units are multiplied or divided, the product or quotient has the units of the resultant quantity (e.g., m and ha are coherent while ft. and acre are not).

The common units of the SI system are the kilogram (mass), the meter (length), the second (time), the Newton (force), the Watt (power), and the Joule (energy). Useful conversion factors include:

Multiply	By	To obtain
Energy, heat, work:		
Btu	2.931×10^{-4}	kWh
Btu	1055	Joule
kWh	3.6×10^{6}	Joule
kWh	3412	Btu
therm (10^5 Btu)	1.055×10^{8}	Joule
toe	4.187×10^{10}	Joule
Power:		
Btu/h	0.2931	Watt
hp	746	Watt
kgf-m/s	9.807	Watt
erg/s	1.000×10^{-7}	Watt
Length:		
Inch	2.54	cm
Foot	0.305	m
Kilometer	0.621	mile
Meter	3.281	ft.
Meter	39.37	in.
Mile	1.609	km
Mile	5280	ft.
Mile	1609	m
Yard	0.9144	m
Area:		
Acre	0.4047	ha
Meter^2	1.0×10^{-4}	ha
Mile^2	2.590	km^2
Volume:		
Foot^3	0.0283	m^3
Gallon (U.S. liquid)	3.785	Liter
Gallon (U.S. liquid)	3.785×10^{-3}	m^3

(Continued)

(Continued)

Multiply	By	To obtain
Barrel (42 U.S. gal)	0.159	m^3
Meter^3	1000	Liter
Mass and density		
Pound (avoirdupois)	0.4536	kg
Metric ton	2205	lbs
Short ton	2000	lbs
Metric ton	1000	kg
Speed:		
Miles/hour	1.609	km/h
Kilometer/hour	0.6214	mph

Energy and GHG equivalencies

Fuel	Btu/gal, lb, ft.^{3}	MJ/kg	GJ/m^3 or MJ/liter
Hydrogen[a]	61,000 Btu/lb	120	0.011
Propane	91,500 Btu/gal	50	26
	21,500 Btu/lb		
Butane	94,670 Btu/gal	45	26
	19,520 Btu/lb		
LPG	90,000−105,000 Btu/gal	50	25−29
	(average 21,500 Btu/lb)		
Natural gas[a]	960−1550 Btu/ft.^{3}	47−56	0.036−0.058
	(average 1000 Btu/ft.^{3})		
Manufactured gas[a]	460−650 Btu/ft.^{3}	29−41	0.017−0.024
	(average 565 Btu/ft.^{3})		
Methane[a]	500−700 Btu/ft.^{3}	28−38	0.019−0.026
Crude oil (42 gal bbl)	5.8×10^6 Btu		38.5
Oil No. 1	136,000 Btu/gal	46	38
	19,800 Btu/lb		
Oil No. 2	138,500 Btu/gal	45	39
	19,400 Btu/lb		
Diesel	130,300 Btu/gal	43	36
	18,400 Btu/lb		

(*Continued*)

(Continued)

Fuel	Btu/gal, lb, ft.³	MJ/kg	GJ/m³ or MJ/liter
Gasoline	127,600 Btu/gal 20,750 Btu/lb	48	36
Kerosene	135,000 Btu/gal 19,810 Btu/lb	46	38
Anthracite coal	12,000–13,000 Btu/lb 24–26 MBtu/ton	28–30	45–48
Bituminous coal	10,000–15,000 Btu/lb 20–30 MBtu/ton	23–35	32–47
Wood (12% moisture)	8000–10,000 Btu/lb	19–23	8–10

Typical information for other woods, ranging from Pines to Live Oak are: specific weight, 28–61 lb/ft.³; energy content, 18–36.6 MBtu/cord or 225–475 kBtu/ft.³.
[a]Note: Data apply at standard temperature and pressure.

Greenhouse gas equivalencies

Energy use form	kg of CO_2 emitted during combustion
Propane (gal)	5.76
Butane (gal)	6.71
Heating oil and diesel (gal)	10.16
Kerosene (gal)	9.75
Jet fuel (gal)	9.57
Aviation gas (gal)	8.35
Gasoline (gal)	8.89
Crude oil (bbl)	430
Natural gas (therm)	5.312
Natural gas (kft.³)	53.12
Coal (short ton):	
Anthracite	2579
Bituminous	2237
Lignite	1266
Average	2100
Asphalt, road oil (gal)	11.95
Lubricants (gal)	10.72
Municipal solid waste (short ton)	2618

Source: U.S. Energy Information Agency (February 2, 2016). *Carbon dioxide emission coefficients.*

APPENDIX 2

The amount of CO_2 in the atmosphere: sources and sinks

The total mean mass of the atmosphere is 5.148×10^{15} metric tons.[1] For about 10,000 years up until the time of the Industrial Revolution, the concentration of carbon dioxide in the atmosphere had not exceeded around 300 ppm on a volumetric basis (ppmv). Following the Industrial Revolution, as the combustion of fossil fuels increased, the concentration of CO_2 began to increase, slowly at first and then more rapidly. After a number of years the annual increase reached 1 ppmv. In a shorter timeframe it soon reached 2 ppmv per year. As of 2018, the concentration of carbon dioxide had increased to 410 ppmv (0.041% by volume). The total mass of CO_2 in the atmosphere can be estimated. To illustrate the method, this is how the preindustrial age mass of CO_2 would be found:

$$0.0280 \text{ vol .\%} \times \left[\frac{44.0095}{28.97}\right] = 0.0425 \text{ mass percent } CO_2,$$

where the molar mass of CO_2 (one C + 2O) = 44.095 g/mole and the mean molar mass of air = 28.97 g/mole (mixture of oxygen, nitrogen, and other atmospheric gases).

From this, the total mass of carbon dioxide in the atmosphere before the Industrial Revolution is:

$0.0425 \times 5.148 \times 10^{15} = \mathbf{2.188 \times 10^{12}}$ metric tons of CO_2, or 2188 Gmt of CO_2

At a concentration of 410 ppmv in 2018, the CO_2 mass in the atmosphere is:

$$0.0410 \text{ vol .\%} \times \left[\frac{44.0095}{28.97}\right] = 0.0623 \text{ mass percent } CO_2,$$

$0.0623 \times 5.148 \times 10^{15} = \mathbf{3.207 \times 10^{12}}$ metric tons of CO_2, or 3207 Gmt of CO_2.

---

[1] David R. Lide, *Handbook of Chemistry and Physics* (Boca Raton, FL: CRC, 1996), 14−1.

At the IPCC recommended limit of 450 ppmv, the mass of CO_2 in the atmosphere can be estimated as: $\mathbf{3.520 \times 10^{12}}$ metric tons of CO_2, or 3520 Gmt of CO_2.

Another useful number to bear in mind is that each additional 7.86 $GmtCO_2$ added to the atmosphere increases the concentration by 1 ppmv, or

$$\mathbf{1\ ppmv = 7.86\ GmtCO_2}$$

Professor James Hansen, in "Climate Change in a Nutshell," December 2018 (see endnotes for citation), estimates that it could cost as much as \$100–\$200 per ton to capture and remove CO_2 from the atmosphere. On this basis, the cost to reduce the atmospheric concentration by 1 ppm would be roughly \$0.8–\$1.6 trillion.

The current excess of greenhouse gases (defined as that amount greater than that present in pre-Industrial Revolution times) is anticipated to remain in the atmosphere for millennia. If the increase in carbon dioxide concentration stops, the average temperature of the earth is not expected to change for hundreds of years.

Prior to the Industrial Revolution the sources and sinks of atmospheric carbon dioxide remained roughly in balance. Carbon dioxide was produced by detritus and by other sources including forest fires, volcanic action, and to a lesser extent by the dissolution of carbonates ("limestone"). Carbon dioxide is removed from the Earth's atmosphere by photosynthesis, absorption in rivers and in the ocean and by the formation of carbonates such as limestone or marble. About 30% of atmospheric carbon dissolves into rivers and oceans, leading to an increase in concentration of carbonic acid.

After the Industrial Revolution, the sources of CO_2 were augmented by human activities. The primary new sources of CO_2 were combustion of coal, petroleum, and natural gas, and the production of cement. Deforestation has been another cause for the increase in atmospheric carbon dioxide, by eliminating plants that would normally absorb CO_2 from the atmosphere. Human-caused forest fires are an additional source. Because these additional sources exceed the amount of carbon dioxide that can be taken up by natural sinks, the concentration of CO_2 in the atmosphere has gradually increased.

APPENDIX 3

Will the IPCC goal of 450 ppm be met?

The IPCC 2014 report suggests that to avoid raising the average global temperature by 2°C (where serious consequences from global warming might occur), it would be prudent to keep the average atmospheric concentration of CO_2 below 450 ppm. (This temperature—2°C—is the average increase compared to the IPCC baseline of 1850–1900.) Recent data show that the average temperature has already increased by 1.0°C, so we are already halfway to the IPCC danger line.

It is helpful to recall that before the Industrial Revolution brought about the growing use of coal, oil, and natural gas, the atmospheric concentration of CO_2 had not exceeded around 300 ppm for thousands of years. It has been rising unusually quickly since then. A milestone atmospheric CO_2 concentration of 400 ppm was reached in November 2015. We compiled the official data for the average CO_2 ppm values for each month from November 2015 to June 2018. As a technical note, the concentration does not increase every month. In each of the 3 years of data we used, the concentration *increases* during October–May, but then *decreases* during June–September. This is attributed to the northern hemisphere growing season, when plants take up CO_2 and slow the amount being released to the atmosphere. Overall, during the 3-year period we examined, the average concentration steadily increased from 400 to 410 ppm (all numbers rounded). In this 32-month period, the concentration rose by 10 ppm. If the assumption is made that the concentration will continue to rise at the *same rate* until it reaches 450 ppm—that is, no human intervention and no sudden acceleration of carbon dioxide releases occurs–the time to reach 450 ppm can be estimated. With these assumptions, from November 2015 (400 ppm) until June 2018 (410 ppm) it took 32 months. At this rate, 450 ppm would be reached sooner than expected.

Here is the discouraging part. We are more than halfway to the 2°C mark. The temperature change thus far, compared to the IPCC 1850–1900 baseline, has been:

- 0.94°C (2015)
- 1.24°C (2016)
- 1.13°C (2017)
- 1.16°C (2018)

These changes amount to anywhere from 0.1°C to 0.3°C per year.

Many scientists are saying that they believe we have passed the point where it is possible to limit the global temperature increase to 2°C. These data seem to confirm that.

This does not bode well for an overheated earth.

APPENDIX 4

Key parameters used to formulate Action Plan

This appendix lists key parameters from estimates used to formulate the Action Plan described in Chapter 14, Action Plan: Efficiency, Power, Transportation, and Land Use. The estimates were developed using an Excel spreadsheet covering the years 2018–2050 and containing 6700 cells. The principal parameters such as 2018 world population, projected 2050 world population, and current greenhouse gas releases were obtained from sources believed to be the most authoritative and current. The sources for data used in the estimates are all listed here so the reader can refer to them for comparison purposes with other sources of information if desired. Note: Only principal results are shown here. Many intermediate steps employing mathematical conversion factors and other adjustments were required in some cases to obtain the final result (Table A.4.1).

Table A.4.1 Key parameters values.

Component	2018	Growth rate	2050	Change
"Business as usual" estimates				
Global population (10^9)	7.6	Variable, average 0.8%	9.9	2.3
Global GDP ($US 10^{12})	84.8	3.25%	236	151.2
Global energy use (10^9 toe/year)	14.3	1.25%	21.3	7.0
Global energy use (10^9 GJ/year)	599	1.25%	891	292
Global electric generating capacity (10^9 Watts)	6473	2.0%	12,199	6461
Global GHG (GmtCO$_{2eq}$)	55	1.25%	81.8	26.8
Global GHG (GmtCO$_2$)	37.1	1.25%	55.2	18.1
Action Plan estimates				
Energy efficiency (ongoing)	1.6%	2.0%		NA

(*Continued*)

(Continued)

Component	2018	Growth rate	2050	Change
Energy efficiency (additional)		1.0%		NA
Global GHG ($GmtCO_{2eq}$) *with 1% efficiency improvement*	55	0.25%	62.6	7.6
Global electricity generation (10^{12} Wh)	26,700	Variable 2.0% declining to 1%	38,700	12,000
CO_2 from global electricity generation (Gmt/year)	11.3	Variable 2.0% declining to 1%	16.4	5.1
50% renewable electricity generation (10^{12} Wh)	6995	Variable 7.9% declining to 0.7%	19,378	12,383
CO_2 saved by 50% renewable electricity generation (Gmt/year)	2.97	Variable 7.9% declining to 0.7%	8.2	5.2
Global cars/light trucks (10^9)	1.2	3.5%	3.0	1.8
Global vehicles ($GmtCO_2$)	3.4	3.5%	8.7	5.3
Global electric vehicles (10^9)	0.0056	Variable 41% declining to 10%	1.1	1.1
Savings in $GmtCO_2$ due to electric vehicles	0.016	Variable 41% declining to 10%	2.53	2.51
Savings in $GmtCO_2$ due to 56 mpg fuel standard for non-EV	0 (no change for 5 years)	Remaining nonelectric vehicles	1.07	1.07
Benefit of additional electrification of 0.5%/year, ($GmtCO_{2eq}$)	0	0.5%/year	3.58	NA
Global forest area (10^9 ha), losing 7.56 10^6 ha/year	4.0	Losing 0.189%/year	3.77	0.23
Added GHG due to deforestation ($GmtCO_{2eq}$)	1.38/year	Carbon capture loss 5 mt/ha/year	1.38/year	
Plant 10^{12} trees (Paris Accord) @31 × 10^9/year saves ($GmtCO_{2eq}$)	0.71	1000 trees/ha; 22.6 kg CO_2/tree	0.71	

NA, Not applicable.

Table A.4.1 sources of information.

1. World population: 2018 World population data 7.6 billion: https://www. prb.org/2018-world-population-data-sheet-with-focus-on-changing-age-structures/.
2. World GDP in 2018 $84.8 billion: (Data from World Bank) http://worldpopulationreview.com/countries/countries-by-gdp/.
3. World Energy Use 14.3×10^9 toe. Used in Chap. 15, p. 181. *Source*: IEA (The International Energy Agency) World Energy Outlook. "Global energy consumption in 2018 increased at twice the average rate since 2010, 2.3%/year to 14,301 Mtoe," https://www.iea.org/weo2018/. See also "BP Energy Outlook, 2019 edition," https://www.bp.com/content/dam/bp/business-sites/en/global/corporate/pdfs/energy-economics/energy-outlook/bp-energy-outlook-2019.pdf.
4. World Energy Use (598×10^9 GJ). Obtained using conversion factor 41.87 GJ x toe. Check: https://www.unitjuggler.com/convert-energy-from-J-to-toe.html.
5. Energy in GJ/3.6×10^6 J/kWh = kWh, or 598×10^{18} J/3.6×10^6 J/kWh = 166×10^{12} kWh. Same growth rate as in items 3 and 4.
6. Average increase in energy/GDP, 1.6%/year average 2000–17: https://yearbook.enerdata.net/total-energy/world-energy-intensity-gdp-data.html. We assume 2.0%.
7. Author's estimate of additional efficiency improvement is 1.0%. Includes ongoing improvements in buildings (LEDs, etc.) and industrial processes. Based on data compiled from Craig B. Smith and Kelly E. Parmenter, *Energy Management Principles* (2nd ed.), Elsevier, 2016.
8. Global electrical generating capacity: 6473 GW in 2018. Assume with business as usual, grow 2.0%/year, reaching 12,199 GW in 2050. Reflects greater use of electricity to replace fossil fuels.
9. Renewable generating capacity (GW). Renewable generating capacity in 2018, 2284×10^3 MW. Forty-six percent was solar/wind. https://www.irena.org/Statistics/View-Data-by-Topic/Capacity-and-Generation/Statistics-Time-Series. Renewable electricity in 2018 had a growth rate of 18%, the 15th year of double-digit growth. https://www.bp.com/en/global/corporate/energy-economics/statistical-review-of-world-energy/renewable-energy.html.
10. Global electricity production in 2018 was 26,700 TWh. https://www.iea.org/geco/electricity/. Electricity will reach 38,700 TWh in 2050. https://about.bnef.com/blog/global-electricity-demand-increase-57-2050/.
11. We assume that nuclear and hydro + coal average out to a heat rate of 8000 Btu/kWh, same as natural gas. With this assumption, we use the carbon yield of natural gas for *all* electricity production. At average 1000 Btu per cubic ft. of natural gas, this is 8 cubic ft./kWh $\times 5.3 \times 10^{-5}$ metric tons/

(Continued)

Table A.4.1 (Continued)

cubic ft. $= 4.24 \times 10^{-4}$ metric tons CO_2/kWh, or 4.24×10^{-13} GmtCO$_2$/kWh.

12. Renewables account for 26.2% of global electricity generation in 2018. https://www.ren21.net/wp-content/uploads/2019/05/gsr_2019_full_report_en.pdf. Renewables now (2018) growing at 7.9%/year. Source: https://www.irena.org/newsroom/pressreleases/2019/Apr/Renewable-Energy-Now-Accounts-for-a-Third-of-Global-Power-Capacity.

13. Note: Climate interactive baseline data for 2018 report 55 Gmt/year CO_{2eq}: https://www.climateinteractive.org/.

14. We assumed land use emissions (LUC emissions) $= 55$ $CO_{2eq}-37.1$ CO_2.

15. Global CO_2 emissions hit new high in 2018 of 37.1 GmtCO$_2$ https://www.scientificamerican.com/article/co2-emissions-reached-an-all-time-high-in-2018/.

16. There are 1.2 billion motor vehicles in the world today, with 2 billion expected by 2035. https://www.greencarreports.com/news/1093560_1-2-billion-vehicles-on-worlds-roads-now-2-billion-by-2035-report. There will be 3 billion light-duty vehicles in 2050. https://www.fuelfreedom.org/cars-in-2050/.

17. At end of 2018, there were 5.6 million battery and plug-in electric vehicle (ev) in the world. The growth rate from 2014 to 2018 was 384,600 to 2.242 million, a factor of 5.8, and an average of 41%/year. https://www.electrive.com/2019/02/11/the-number-of-evs-climbs-to-5-6-million-worldwide/. Estimates of ev (and hydrogen) range from 1 to 1.25 billion in 2050. https://www.fuelfreedom.org/cars-in-2050/.

18. Building efficiency improvements are included in 1% efficiency improvement (item 7).

19. Forests cover 30% of the earth's surface (4 billion ha). http://www.earth-policy.org/indicators/C56/forests_2012. Loss of forests contributes between 12% and 17% of annual GHG emissions. https://www.conserve-energy-future.com/various-deforestation-facts.php. Earth losing 18.7 million acres/year (7.56 million ha/year). This reference says 15% of GHG from deforestation. https://www.livescience.com/27692-deforestation.html.

20. Agriculture: Arable land and land under permanent crops is currently 13.4 billion ha. This is 36% of land suitable for farming that is presently used. http://www.fao.org/3/y4252e/y4252e06.htm. About 80% is used for livestock. https://ourworldindata.org/agricultural-land-by-global-diets. About 9% of GHG emissions come from agriculture. https://www.epa.gov/ghgemissions/sources-greenhouse-gas-emissions. Other estimates say up to 33%. https://www.nature.com/news/one-third-of-our-greenhouse-gas-emissions-come-from-agriculture-1.11708. Regarding future food and agricultural land requirements, see The World Resources Institute, Executive Summary Synthesis Report. https://www.wri.org/our-work/topics/food.

(Continued)

Key parameters used to formulate Action Plan 283

Table A.4.1 (Continued)

21. Reforestation benefits in $GmtCO_2$/year: Tropical forests, 1.91, temperate forests, 0.71, reduced firewood use, 0.49. Total, 32 years, 99.52. These estimates from *Drawdown*.

22. Gains from sequestration: Meeting commitment of Paris Agreement requires planting 1 trillion trees by 2050. Trees sequester between 20 and 200 kg per tree. We assumed 22.6 kg (50 lbs)/tree. See https://projects.ncsu.edu/project/treesofstrength/treefact.htm (Says 48 lbs/tree in the United States) and http://www.truevaluemetrics.org/DBpdfs/Forests/Tree-Nation-Tropical-tree-sequestration-of-CO2.pdf for discussion of tropical trees.

APPENDIX 5

Flood and sea rise mitigation

Flooding due to storms and sea rise is of great concern to coastal inhabitants. It is estimated that millions of persons throughout the world live at elevations of 3 ft. or less above sea level and are at risk.[1]

In Newport Beach, California, the authors' hometown, the City has taken the first tentative steps toward addressing sea level rise. Under California law, the City has had to include a sea level rise element in its master plan. The City has to specify how it will cope with a projected 7-foot rise in sea level by 2100. Among the measures currently being implemented are raising foundation levels by 3 ft. for new homes constructed on the islands in Newport Harbor or other exposed coastal areas. In addition, portions of the seawall surrounding Balboa Island are being raised 9 in., bringing it to a height of 10 ft. above mean sea level. This is a short-term or stop-gap measure, since at this height, there have already been times when the seawalls elsewhere have been overtopped on days when a king tide coincided with a storm surge.

Seawalls are expensive. Balboa Island's 2 miles of 80 + -year-old seawalls are near the end of their useful life. Replacing them has a price tag of $68 million. This is a burden the City Council did not want to face when it is already concerned about unfunded pension costs. Instead, the city opted for a reduced plan, adding a 9 in. cap to the existing seawalls, at a cost estimated to be between $1.5 and $2 million. The City's public works department admits it is a temporary measure and will prove inadequate after a decade or so if the sea continues to rise. In addition, for planning purposes, the City admitted that it used a design based on the lower estimate of sea level rise out of budgetary concerns.

During the winter months when storms are likely to bring high swells to Newport's beaches, the City creates a sand berm several feet high along the beachfront to keep seawater from flooding into beachfront homes and the Balboa Village commercial area. This has not prevented occasional flooding, which will only get worse as the sea level rises.

---

[1] Goodell, *The Water Will Come*, p. 69.

Further south along the coast, the city of Del Mar is adopting a more draconian approach. Here there are expensive beachfront homes that are 3 ft. above mean sea level. Del Mar is proposing to adopt a *managed retreat* plan whereby residents in the threatened areas will be encouraged to vacate their homes so they can be torn down, creating a buffer zone. This obviously raises serious financial issues. Who pays? Also, if the city adopts such a plan, will it have the immediate effect of lowering property values and making it harder to sell the properties? These are complex issues, with no good choices.

Author Jeff Goodell writes, "The water will come." It is not "if" but "when."

In the South Pacific, the idea of managed retreat is already a reality. Faced with the immediacy of rising seas that have flooded low-lying areas and destroyed homes and farms, the government of Kiribati, a group of islands in Micronesia, has purchased 5500 acres of land on high ground on an island in the Fiji islands for $8.8 million. This property will become the new home for those displaced by sea level rise. The government of Fiji has indicated that it will accept these climate refugees.

Alaska is another place where managed retreat has already begun. A combination of rising seas, beach erosion, and subsidence (due to the melting of permafrost) is making villages uninhabitable. Residents are packing up and moving inland to higher ground. According to the U.S. Army Corps of Engineers, 31 Alaskan communities face "imminent" threats from coastal flooding and erosion. For the residents of the island of Shismaret, rising seas are not a future concern but a present reality. The residents have voted to leave the island and move to the mainland. However, at this point there is no source of funding to make the move.[2]

If you asked, they will say, "The seas have come."

In California there is $100 billion of infrastructure that will be exposed by 1.4 m (4.6 ft.) of sea level rise. Two-thirds of California's beaches will be eroded by the year 2100. The year 2017 set a record of $300 billion in flood damages. There were 15 different incidents across the U.S. with costs exceeding $1 billion.[3]

One is moved to wonder how to balance the expense of short-term protective measures versus the cost of a full-fledged commitment toward

---

[2] Oliver Milman, "Alaskan Towns at Risk from Rising Seas Sound Alarm as Trump Pulls Federal Help," *The Guardian*, August 10, 2017.

[3] Brett Sanders, University of California, Irvine (2018), in a presentation at "Speak-up Newport." Newport Beach, CA.

fighting global warming? What if every coastal city in the United States put pressure on the U.S. House of Representatives at the Federal government level to take meaningful action on global warming, rather than denying it was a problem?

When will residents of coastal cities decide that hunkered down behind 6-foot high seawalls with no view of the ocean, is not the lifestyle they imagined? Will residents of Balboa Island, Del Mar, California, and Miami, Florida accept living in homes built on 10-foot high stilts?

We need to think about it. The water is coming. (With thanks to Jeff Goodell)

APPENDIX 6

Financial measures

Cap and Trade

In 2006 California's legislature passed AB 32, "*The global warming solutions act of 2006*," that established a series of measures to lower California's greenhouse gas (GHG) emissions to 1990 levels by 2020. "Cap and Trade" is a key element of the program to ensure that the goal is met. The bill was recently extended for another 10 years, to 2030. The initial objective was to reduce California's GHG emissions to 431 million metric tons of CO_2 by 2020 from a baseline of 509 million metric tons. California's goal was to reduce GHG emissions by 2.6 million tons of CO_2/year, or 0.0026 Gt/year. This goal was met ahead of schedule in 2016. New legislation extended the plan to 2030, with a new goal to reduce GHG by another 40% below the 2020 goal, or to 259 million metric tons of CO_{2eq} per year.

Besides growth in renewable energy and more efficient vehicles, California is using Cap and Trade to reduce GHG. Each year a "cap" or limit is set on the amount of GHG that can be released by any source, such as an electric utility or refinery. Initially the cap was lowered by 3% each year, to ensure that the 2020 goal was met. If a polluting source cannot meet its "cap," it can buy "allowances" at a state auction. Each allowance allows it to emit one metric ton of CO_2 equivalent. Allowances are provided by entities that invest in GHG reduction technologies (e.g., solar energy) or by state programs that will reduce GHG.

In January 2018, California and the Province of Quebec formed a joint Cap and Trade program that created the largest carbon fee market in North America. Recently the mean price per metric ton of CO_2 (August 2018 auction) was $15.80. The auction raised in excess of $79 million that California and Quebec will invest in GHG reduction projects.

Funds raised by Cap and Trade, along with other contributions from the state budget, are administered by the California Climate Investment organization. Funds are used to improve public health, encourage sustainable communities, reduce transportation GHG emissions, and support other approved initiatives.[1]

Fee and dividend

In November 2018, the U.S. House of Representatives introduced a bipartisan bill titled the *Energy Innovation and Carbon Dividend Act* (HR763). In 2019, the bill had bipartisan support with over 60 cosponsors. The bill recognizes that human activities are causing rising global temperatures, and has as its goal to reduce U.S. emissions by 33% within 10 years and 90% by 2050 (compared to 2015) by imposing a carbon fee. It offsets the impact of higher energy cost on low-income families by returning a pro rata share of the net income (after administrative costs) to individual taxpayers and 2 children under 19 years of age.

Justification for the proposed legislation is based on the risks to human health, the natural environment, and global civilization if temperatures continue to increase unchecked.

- Benefits: the measures proposed in this legislation will benefit the economy by creating 2.1 million additional jobs over the next 10 years. Additional savings will come from lower health-care costs and preventing 13,000 pollution-related U.S. deaths annually.
- Additional benefits of phasing in carbon fees on GHG emissions are: (1) it is the most efficient, transparent, and enforceable mechanism to drive an effective and fair transition to a domestic energy economy; (2) it will stimulate investment in alternative energy technologies; and (3) it gives all businesses powerful incentives to increase their energy efficiency and reduce their carbon footprints in order to remain competitive.
- The "dividend" (or rebate) from carbon fees will be paid to every American household to help ensure that families and individuals can afford the energy they need during the transition to a GHG-free economy. The dividends will stimulate the economy and remove any burden due to higher prices.

---

[1] *California's 2017 Climate Change Scoping Plan* (Sacramento, CA: California Air Resources Board, 2017), 132 pp., https://www.arb.ca.gov/cc/scopingplan/scoping_plan_2017.pdf.

This legislation would impose a carbon fee on all fossil fuels and other GHG at the point where they first enter the economy. The fee will be collected by the Treasury Department. Initially the fee will be $15 per metric CO_2 emitted. The Department of Energy will propose and promulgate regulations setting forth CO_2 fees for other GHG including at a minimum methane, NOx, sulfur hexafluorides, hydrofluorocarbons, and perfluorocarbons. The Treasury Department will collect the fees imposed upon all GHG sources and place them in a Carbon Fees Trust Fund for rebate to American households. Each year the increase in carbon fee will be at least $10 per ton of CO_2 unless the Department of Energy determines that a greater increase is required to meet the goal of U.S. CO_{2eq} emissions in 2050 equal to or less than 10% of the 2015 amount.

In order to ensure there is no domestic or international incentive to relocate production of goods or services to regimes more permissive of GHG emissions, and thus encourage lower global emissions, a carbon fee tariff shall be charged for goods entering the United States from countries without comparable carbon fees/carbon pricing. Carbon fee rebates shall be used to reduce the price of exports to such countries. The State Department would determine rebate amounts and exemptions if any.

Other countries and jurisdictions—40 in all—have already proposed or enacted legislation that prices carbon. The total is divided roughly 50/50 between countries with Cap and Trade and those with a carbon tax. Canada recently passed a carbon tax law that starts at C$10/mt and increases by C$10/mt each year up to C$50/mt in 2022. Elsewhere, carbon taxes range from $1/mt in Mexico to a high of $139/mt in Sweden, with the majority between $10/mt and $25/mt.[2]

As our case study of the effect of the 1973 oil embargo demonstrates, there is no doubt that increasing the cost of fossil fuels by tax or other methods is effective in reducing fuel use and stimulating more efficient alternatives. Too rapid a change can damage the economy. Any increase in price needs to be gradual.

---

[2] Matthew Carr, "Carbon Pricing," Bloomberg QuickTake, *Bloomberg Business Week*, Issue #10, January-June 2019, pp. 40—41.

APPENDIX 7

Activist and lobbying groups, litigation examples

Activist groups

- Fridays for Future (FFF)[1,2,3]

   In August 2018, a 15-year-old Swedish student named Greta Thunberg decided that her government was not taking strong enough action to address global warming. She ditched school and sat in front of the Swedish Parliament building holding a handmade sign every school day for 3 weeks to protest the government's failure to act. She sent Instagram and Twitter messages about what she was doing. Within days, her messages went global.

   On September 8, Greta decided to continue striking every Friday until Swedish policies provided a safe pathway well under the 2°C limit recommended by the Paris Agreement. Because of Greta, the hashtags #Fridaysforfuture and #climatestrike spread and students and adults around the world began to protest outside of their parliaments or local city halls.

   Today the FFF website lists the websites for FFF organizations in 25 countries. It has a world map that shows hundreds of cities that have weekly student events. The website reported that on May 24, 2019, student strikes took place in over 1850 cities in 131 countries. The number of student participants was estimated to be well above 1 million.

   In January 2019, Thunberg attended the World Economic Forum's annual meeting in Davos, Switzerland. There the 16-year-old spoke to a large group of millionaires and billionaires. In addressing a panel, she said, "Some people, some companies, some decision-makers in

---

[1] Erik Kirschbaum, "Climate Rallies Draw Thousands in Germany," *Los Angeles Times*, March 30, 2019, p. A3.
[2] Mark Hertsgaard, "The Climate Kids Are Coming," *The Nation*, March 25, 2019, p. 19.
[3] Erik Kirschbaum, "Fridays Are for Climate Change Rallies," *Los Angeles Times*, March 29, 2019, p. A3.

particular have known exactly what priceless values they have been sacrificing to continue making unimaginable amounts of money." She paused, then said, "I think many of you here today belong to that group of people."[4]

At the March 15, 2019, strike in London, a group of students calling themselves the "Student Climate Network," issued a document titled "Manifesto for tackling the climate change crisis." It demands governments to declare that the world faces a climate emergency and they should prioritize protection of life on earth. Other demands included better education, bringing the climate crisis into the class-room, providing the public with the necessary facts, and lowering the voting age to 16![5]

From Santa Fe New Mexico, an 18-year-old student named Hannah L. Abram wrote:

> *We are living in the sixth mass extinction. Ice is melting. Forests are burning. Waters are rising. And we do not even speak of it. Why?*
>
> *Because admitting the facts means admitting crimes of epic proportions by living our daily lives. Because counting the losses means being overpowered by grief. Because allowing the scale of the crisis means facing the fear of swiftly impending disaster and the fact that our entire system must change. But now is not the time to ignore science in order to save our feelings. It is time to be terrified, enraged, heartbroken, grief-stricken, radical.*
>
> *It is time to act.[6]*

The next event was the World Climate Strike, timed to coincide with the UN Climate Action Summit, September 23, 2019. Organizers stated that 7.6 million people took part in 4500 locations in 150 countries. In a fiery speech, Thunberg addressed the United Nations, saying, "We'll never forgive you if action is not taken. . . . . . and we will be watching."

- Extinction Rebellion (XR)

  XR is an international movement that uses nonviolent civil disobedience to achieve radical change in order to minimize the risk of human extinction and ecological collapse. It was born on October 31,

---

[4] Hertsgaard, *The Nation*, March 25, 2019.

[5] "A manifesto for tackling the climate change crisis," *The Guardian*, March 15, 2019.

[6] Jessica Glenza, "Climate Strikes Held Around the World — As it Happened," *The Guardian*, March 15, 2019, https://www.theguardian.com/environment/live/2019/mar/15/climate-strikes-2019-live-latest-climate-change-global-warming.

2018, when 1500 people gathered on Parliament Square in London to announce a "Declaration of rebellion against the UK government."

Several weeks following this initial protest, "6000 protesters converged on London to peacefully block five major bridges across the Thames. We planted trees in the middle of Parliament Square and dug a hole there to bury a coffin representing our future. We superglued ourselves to the gates of Buckingham palace as we read a letter to the Queen. Our actions generated huge national and international news and, as news spread, our ideas connected with tens of thousands of people around the world."[7]

At the core of XR's philosophy is nonviolent civil disobedience, which is believed to be necessary to bring about change. The group does not focus on traditional approaches such as petitions or writing to politicians, but is more likely to take risks (e.g., arrest or jail time). It proclaims, "We want to promote civil disobedience, in full public view—to shake the current political system and to raise awareness. We believe that it is only through this approach that real change can take place."

The movement has rapidly gained momentum, as its declaration of principles was signed by nearly 100 intellectuals, scientists, and public figures. Besides the United Kingdom, XR has chapters in more than a dozen other European countries, South Africa, and the United States. Are the student movements making an impact? Given the global publicity that has followed the student's climate strike and XR, it appears that their tactics are working. Public concern over the climate crisis has reached new levels. As Forbes magazine wrote in a recent article, "If you have not heard of Extinction Revolution yet, you soon will."[8]

By giving talks to local communities across the country, XR has humanized the climate crisis. Instead of recounting the changes wrought by global warming, such as sea level rise, spread of diseases, and crop failure, it is emphasizing that this crisis is about our children. It expresses grief for children, wildlife, for nature, and for fear at the degrading of the systems that keep humans alive. In XR's viewpoint, too many people in positions of power believe that if they deny that this situation we face is serious, then it will not be. They cannot see

---

[7] Extinction Rebellion (XR) website. See https://rebellion.earth/.

[8] Enrique Dans, "If You Haven't Heard of Extinction Revolution Yet, You Soon Will," *Forbes*, April 21, 2019, https://www.forbes.com/sites/enriquedans/2019/04/21/if-you-havent-heard-of-extinction-rebellion-yet-you-soon-will/#34905805396b.

that the disruption we have now is nothing compared with the coming fury from nations of parents realizing that despite certain science, science has been ignored and the worst-case scenario may come to pass, leaving their children facing a terrible future.[9]

- Citizens' Climate Lobby (CCL)[10]

CCL is a nonprofit, nonpartisan, grassroots advocacy organization focused on national policies to address climate change. While CCL is primarily concerned with the United States, it has an international membership. CCL operates with a respectful, nonpartisan approach to climate education and is designed to create a broad, sustainable foundation for climate action across all geographic regions and political inclinations. By building upon shared values rather than partisan divides and empowering its supporters to work in keeping with the concerns of their local communities, CCL works toward the adoption of fair, effective, and sustainable climate change solutions. The main thrust of CCL is to lobby the U.S. Congress to enact a carbon fee and dividend law. The CCL proposal has four main points:

1. Collection of Carbon Fees/Carbon Fee Trust Fund: Upon enactment, impose a carbon fee on all fossil fuels and other greenhouse gases at the point where they first enter the economy. The fee shall be collected by the Treasury Department. The fee on that date shall be \$15 per ton of CO_2 equivalent emissions and result in equal charges for each ton of CO_2 equivalent emissions potential in each type of fuel or greenhouse gas. The Department of Energy shall propose and promulgate regulations setting forth CO_2 equivalent fees for other greenhouse gases including at a minimum methane, nitrous oxide, sulfur hexafluoride, hydrofluorocarbons, perfluorocarbons, and nitrogen trifluoride. The Treasury shall also collect the fees imposed upon the other greenhouse gases. All fees are to be placed in the Carbon Fees Trust Fund and rebated to American households as outlined in #3 below.

2. Emissions Reduction Targets: To align US emissions with the physical constraints identified by the Intergovernmental Panel on Climate Change (IPCC) to avoid irreversible climate change, the yearly increase in carbon fees including other greenhouse gases, shall be at least \$10 per ton of CO_2 equivalent each year. Annually, the Department of Energy shall determine whether an increase larger than

---

[9] Extinction Rebellion (XR) website. See https://rebellion.earth/.

[10] This write up is extracted from CCL's website at: https://citizensclimatelobby.org/carbon-fee-and-dividend/.

$10 per ton per year is needed to achieve program goals. Yearly price increases of at least $10 per year shall continue until total U.S. CO_2-equivalent emissions have been reduced to 10% of U.S. CO_2-equivalent emissions in 1990.

3. Equal Per-Person Monthly Dividend Payments: Equal monthly per-person dividend payments shall be made to all American households (1/2 payment per child under 18 years old, with a limit of two children per family) each month. The total value of all monthly dividend payments shall represent 100% of the net carbon fees collected per month.

4. Border Adjustments: In order to ensure there is no domestic or international incentive to relocate production of goods or services to regimes more permissive of greenhouse gas emissions, and thus encourage lower global emissions, Carbon-Fee-Equivalent Tariffs shall be charged for goods entering the U.S. from countries without comparable Carbon Fees/Carbon Pricing. Carbon-Fee Equivalent Rebates shall be used to reduce the price of exports to such countries. The State Department will determine rebate amounts and exemptions if any.

In 2019, CCL is backing HR763, the bipartisan House of Representatives bill titled *Energy Innovation and Carbon Dividend Act*.

- John Kerry's Initiative[11]

Former Secretary of State (Obama administration) John Kerry is the person who negotiated and signed the Paris Agreement on behalf of the United States. Embittered by the failure of the Trump administration to exercise leadership in this vital international agreement, Kerry announced that he is working with former U.S. Energy Secretary Ernest Moniz and others to develop an international initiative to hold politicians accountable for undermining the fight against climate change. The goal is to "punish President Trump and other politicians for failing to combat global warming." The details of the new "climate accountability initiative" are not yet final at the time of this writing but will be announced in the near future.

This will be interesting. One wonders if there is some legal principle whereby political leaders who deny global warming or continue to encourage greenhouse gas emissions, or fail to conform to international agreements could be brought before the International Criminal Court (ICC), Den Hague, The Netherlands, for crimes against humanity. If so, such an initiative would appeal to many.

---

[11] Tony Barboza, "John Kerry is on a Climate Mission," *Los Angeles Times*, April 15, 2019, p. B3.

Litigation

- City and County lawsuits
  The lawsuits began when *Inside Climate News* and the *Los Angeles Times* reported that internal documents prove that for 30 years ExxonMobil "understood the science of global warming, predicted its catastrophic consequences, and then spent millions to promote misinformation."[12]
  A growing number of cities and counties, from New York to San Francisco, have sued major oil companies requesting compensation for climate change damages. These include Richmond, California, where a large Chevron refinery is located, San Mateo County, Marin County, Imperial Beach County, San Francisco, Oakland, Santa Cruz and Santa Cruz County, New York City, Boulder Colorado, Boulder County, San Miguel County Colorado, and King County Washington, to name a few. In some cases, the suits seek billions of dollars to reimburse the cost of constructing seawalls and other mitigation expenses. Most of the lawsuits are ongoing, while being vigorously contested by ExxonMobil, Shell, BP, Chevron, and ConocoPhillips. Some lawsuits have been dismissed when the judge ruled that global warming dangers are real but the issue should be resolved by Congress. Some dismissed cases are currently undergoing appeals.
- State lawsuits
  The Attorneys General of New York, Massachusetts and the U.S. Virgin Islands have initiated investigations into ExxonMobil to determine if the company lied to the public about what it knew about climate change, or to shareholders about how climate risk might devalue the company. ExxonMobil raised a vigorous and costly fight against the investigations, forcing the Virgin Islands to quit within a few months, while New York and Massachusetts continued the court fight. Despite a series of delaying maneuvers, the cases continue to make their way through the courts. In 2018, the state of Rhode Island also filed a lawsuit against fossil fuel companies.
- Children's lawsuits
  In 2015, an organization called "Our Children's Trust" sued the U.S. federal government on behalf of 21 children and young adults.

---

[12] David Hasemeyer, "Fossil Fuels on Trial: Where the Major Climate Change Lawsuits Stand Today," *Inside Climate News*, January 6, 2019.

Scientist James Hansen's granddaughter is among the 21 plaintiffs. The children's lawyer, Julia Olson from Our Children's Trust, filed the lawsuit in Oregon. It is now known as *Juliana versus United States*. In the suit, the children accused the government of violating their rights under the Fifth Amendment of the Constitution by promoting the use of fossil fuels and failing to take action to prevent global warming. They make the point that it is their generation that will suffer from the government's actions.

The lawsuit is based on a unique principle, called the "public trust doctrine." The public trust doctrine is a legal concept that says that the government holds resources such as water, land, and air in trust for its citizens. On this principle, the government has failed in its duty to restrict fossil fuel uses and control greenhouse gas emissions, despite knowing that the growing use of fossil fuels was altering the climate.

After *Juliana versus United States* was filed, the Obama administration and then the Trump administration filed numerous requests to dismiss or otherwise halt the case. To date these efforts have failed to stop the case from proceeding through the courts. Since the original filing of *Juliana versus United States*, Our Children's Trust has supported the filing of nine similar cases in state courts from Alaska to Florida. Judges in the Alaska, Oregon, and Washington cases dismissed them, but the Alaska case is being appealed.

In a recent Oregon hearing of *Juliana versus United States*, Judge Andrew Hurwitz told attorney Olson that she presented "compelling evidence that we have an action by the other two branches of the government. It may even rise to the level of criminal neglect."[13] In their case, the children argue for changes in government efforts to control greenhouse gas emissions and for stopping programs that subsidize or encourage the development and use of fossil fuels. In addition, the suit asked the court to issue an injunction to stop the federal government from issuing new leases for coal mining and offshore oil and gas projects.[14]

Their case has received widespread support from physicians, health organizations, and two former U.S. surgeons general. The surgeons general stated that "(the children) were born into this problem; they did not create it."

On January 17, 2020, The Ninth U.S. Circuit Court of Appeals rejected the case, ruling: "The panel reluctantly concluded that the plaintiffs' case must be made to the political branches or to the electorate at large."[15] The case is still pending in several state courts.

---

[13] Nina Pullano, "Kids Face Rising Health Risks from Climate Change, Doctors Warn as Juliana Case Returns to Court," *Inside Climate News*, June 4, 2019.

[14] Julia Rosen, "Youths Sue for a Livable Climate," *Los Angeles Times*, June 3, 2019, p. A1.

[15] Brent Kendall, "Climate-Change Suit is Thrown Out," *The Wall Street Journal*, January 18—19, 2020.

APPENDIX 8

Excerpts from corporate annual reports

Bank of America

Bank of America announced plans for its operations to become carbon neutral and purchase 100% renewable electricity by 2020. 2010–20 operational goals included achieve carbon neutrality, reduce energy use by 40%, and purchase 100% renewable electricity. In addition, in 2015 the bank announced its second environmental business initiative to provide $125 billion by 2025 for supporting low-carbon and sustainable business through lending.

Olin Corporation

Electricity is a predominant energy source for Olin's manufacturing facilities with approximately 76% of electricity generated from natural gas or hydro.

Eli Lilly Corporation

Using 2012 as a baseline, Eli Lilly aims for a 20% reduction in greenhouse gas emissions by 2020 and 20% improvement in energy efficiency by that time.

Verizon

The 2016 annual report discussed a stockholder proposal requesting that Verizon adopt a science-based greenhouse gas reduction target consistent with the IPCC 2°C scenario. The board recommended a vote against this proposal stating as a reason that "a connected world is a more sustainable world." In addition, they stated that they had a goal to reduce carbon intensity by 50% over 2009 baseline by 2020. They reported that that goal was met in 2016, by installing economizers to make cooling systems more efficient, migrating from copper to optical fiber, and making building improvements and changes in fleet operations and investing in renewable sources of energy. They stated that Verizon generates over 24 MW of green energy in the United States and has a goal of doubling it by implementing an additional 24 MW by 2025.

The Southern Company

Shareholders proposed the same measure as stated above for Verizon for the Southern Company. Again the board recommended voting against this measure, citing its intent to continue with a full portfolio of energy resources, natural gas, coal, nuclear, and renewables. It is proceeding with the construction of the only new nuclear power plant being built in the United States, Vogtle units 3 and 4. In Kemper County, Mississippi, it operates a facility that will convert Mississippi lignite coal into "clean-burning" synthesis gas for reducing emissions of sulfur dioxide, nitrogen oxides, carbon dioxide, and mercury. The company claims to have invested $2 billion in wind projects in 2016 with 1400 MW generating capacity.

DuPont Corporation

DuPont reports having reduced greenhouse gas emissions by 7% between 2010 and 2015 and also reducing nonrenewable energy intensity by 3.3%.

Excerpts from corporate annual reports   303

Caterpillar

Caterpillar cites three energy-related goals for 2020: reduce energy intensity by 50% compared to 2006, use alternatives/renewable sources to meet 20% of Caterpillar's energy needs, and reduce greenhouse gas emissions by 50% compared to 2006.

Home Depot

Home Depot's goal is to procure 135 MW of renewable energy from various sources by the end of 2020. In fiscal 2016 the company announced an investment in wind-powered renewable energy that will provide it with 50 MW—enough to power 100 Home Depot stores.

Chevron Corporation

A shareholder proposal focused on Chevron's risk profile, stating that the company's historic capital spending on high-cost, high-carbon assets has eroded profitability and made the company vulnerable to a downturn in demand and subsequent fall in oil prices. The proposal requested that Chevron prepare a plan on how the company will transition to a low-carbon economy. The board rejected this proposal, stating that it supported mitigation of greenhouse gas emissions, adaptation to climate change, and continuation of scientific and technological research, but gave no specifics.

Alliant Energy

Since 2010, Alliant has retired, repurposed, or converted one-third of its coal-fired generation and is continuing with highly efficient natural gas–fired generation facilities. The company is also making a $1 billion investment in Iowa and wants to add 500 MW of wind energy by the end of 2020. The company operates one of Wisconsin's largest solar facilities in Rock County and supports energy efficiency assistance to customers to improve operations.

DTE Energy Inc

The annual report notes that DTE owns or previously owned a number of manufactured gas plant sites that are contaminated. In addition it has coal ash storage facilities. All of these are subject to regulations that require cleanup and represent a future liability. In the annual report DTE states that it will be expected to comply with new Michigan legislation increasing the percentage of power required by renewable energy sources. It cannot predict the financial impact and does not outline any plan to comply with these regulations. In its 2017 annual report, a shareholder proposal asked the company to publish an assessment of how it would comply with meeting the IPCC warming limit of $2°C$. It should be noted that DTE Energy is the 17th largest CO_2 emitter in the United States and relies on coal for 70% of its power generation. The board rejected this proposal, noting that the company prepares an annual "corporate citizenship report" which states a goal to reduce carbon dioxide emissions by 20% below 2010 levels by 2020 and a 40% reduction by 2030.

UPS Corporation

UPS pursues sustainable business practices worldwide through improving efficiency of operations, fleet, and facility engineering projects. Using 2007 as a baseline, UPS established a companywide target to reduce carbon intensity by 20% by 2020. As of 2015, the company stated it had achieved a 14.5% reduction of greenhouse gas emissions.

Wisconsin Electric (WEC Energy Group)

Wisconsin Electric has a goal of reducing CO_2 emissions from its electric generating fleet by approximately 40% below 2005 levels by 2030.

Excerpts from corporate annual reports 305

Eaton Corporation

Eaton reported that in 2016 it reduced the total amount of greenhouse gases emitted by its operations by 51,000 metric tons. In the same period it also carried out 53 energy efficiency projects that eliminated an additional 3000 metric tons of greenhouse gases.

Duke Power

Duke reports ongoing investments in clean energy facilities, more natural gas generation, and renewables to move toward a low-carbon future. This includes the retirement of more than 40 old coal units and has led to 29% reduction in carbon dioxide emissions since 2005. In 2016 the company added 400 MW of wind and 150 MW of solar. The company projects that with these investments by 2026 it will have reduced carbon emissions by 35% from 2005 levels.

Corning Inc

Under "risk factors" in its annual report, Corning makes a statement that it has "taken steps to control the amount of greenhouse gases created by our manufacturing operations. However, we cannot provide assurances that environmental claims will not be brought against us or the government regulators will take steps to adopt more stringent environmental standards."

Index

Note: Page numbers followed by "*f*," "*t*," and "*b*" refer to figures, tables, and boxes, respectively.

A

Acidification, ocean, 45
Action Plan, 201, 204*b*, 213*t*, 216*t*, 239
  business as usual, 203, 214*f*
  carbon tax/fee, 211–212, 223–226, 225*t*
  challenges of global approach, 202–203
  effective CO_2 capture systems development, 212
  energy efficiency and conservation, 204, 206*f*
  heat and power, renewables for, 207–208
  key parameters
    sources of information, 281*t*
    values, 279–284
  land-use changes and agriculture improvement, 211
  mitigation, 219–223, 221*f*
    measures, 220–222
    sea wall construction, United States, 220
  moon shot, 201–202
  nuclear power, 212
  petroleum-based transportation fuels, 208
  replace coal by solar and wind, 207
  silver bullets, 218–219
  stop deforestation and replant cleared areas, 209, 210*f*
Actions for concerned citizens
  changing political landscape, aware of, 252
  divest from major carbon emitters, 251–252
  educating the public, 250
  engage with climate change activists, 250–251
  individual actions, 252
  litigation, 251
  lobby for government action, 251
Actions for industry
  changing corporate philosophies, 254
  proactive about fighting global warming, companies, 254–255
  corporate naysayers, 255–256
  keep innovating, 253–254
  small group of bad actors, 256
Activist groups, 293–297
  Citizens' Climate Lobby (CCL), 296
    border adjustments, 297
    collection of carbon fees/Carbon Fees Trust Fund, 296
    emissions reduction targets, 296–297
    equal per-person monthly dividend payments, 297
  Extinction Rebellion (XR), 294–296
  Friday's for Future (FFF), 293–294
  John Kerry's Initiative, 297
Aerosol sprays, 185
Afforestation, 209
Agriculture, 90–92, 170–171, 211
Air conditioning, efficiency improvements, 3, 253–254
Air pollution, 4, 79, 235
Air storage system, 142
Alaska, 46, 155, 182, 219, 254, 286, 299
Alaskan Center for Climate Assessment and Policy, 46
Albania, 248
Albedo, 77, 218–219
Alliance of Small Island States (AOSIS), 104
Alliant energy, 303–304
Alternating current (AC), 142
Aluminum, manufacturing, 62
Amazon, deforestation, 120–121, 209
Animal species, migration, 6
Antarctic, 45, 84, 185
Antarctica ice sheet, 48

307

Index

Apollo program, 202, 260
Arab members of OPEC (OAPEC),
181−182
Arab Oil Embargo, 36
Arab Spring, 183
Aragonite, 83−84
Arctic, 98
  global warming, 45−46
  polar bears, 84
  sea ice, 50*f*, 51*f*
Argentina, 51
Arizona, 102, 145, 194, 247
Arrhenius, Svante, 39
Asteroid, 22−23
Asylum, political, 96−97
Atlantic Ocean, CO_2 in, 45
Atmospheric CO_2 concentration, increase
  in, 41−42
Atomic bomb test, 259
Atomic weight, 68
Australia, 48, 91, 140, 145, 209−210,
  254−255
Australian National Tide Facility, 81
Austria, 53
Automobile
  electric vehicles, 208, 248
  emissions, 184−185
  mileage standards, 184
Aviation fuel, 3, 137, 172
Axis Powers, 202

B

Bakersfield, 155−156
Balboa Island, 82, 285, 287
Bangkok, 80
Bangladesh, 14, 80, 220
Bank of America, 301
Battery technology, 138−139
Beachfront
  erosion, 82
  protection, 285−286
Beef cattle, methane emission, 211
Berkshire Hathaway, 223−224, 256−257
Biodiesel, 135, 209
Biofuels, 115, 129, 133−134, 137, 152,
  172
  advantages, 136*t*

biodiesel, 135
biogas, 135
charcoal, 135
corn-based ethanol, 134
disadvantages, 136*t*
feedstock, 134
first-generation, 134
fossil fuels energy content, compared to,
  135*t*
life cycle, 134*f*
Biogas, 135
Biological feedstock, 133−134
Biosphere, 13, 117−118
Black carbon fuel, 135
Bloomberg News, 139
Bloomberg New Energy Finance (BNEF),
  118
Border adjustments, CCL, 297
Bougainville, 81−82
Bramble Cay melomys, 89
Brazil, 35*t*, 134, 195−196
British Antarctic Survey, 185
British Petroleum (BP), 256−257
Brown, Jerry, California Governor, 102
Buffet, Warren, 223−224
Bullet Train, California, 197
Business as usual, 116−117, 116*f*,
  203−204, 214, 214*f*, 216

C

Cabo San Lucas, 57
Calcium carbonate, 27, 83−84
California Air Resources Board, 184
California case history, 58−60
California Energy Commission, 146
California Environmental Protection
  Agency, 58−60
Calving, 48
Cancer, 186−187
Cap and Trade program, 156, 176, 225,
  289−290
Capital cost, 142−143, 228
Carbohydrates, 69
Carbon-12 (C-12), 69
Carbon-14 (C-14), 68−70
  history, 69
  Libby's discovery, 69

Index

Carbon budget, 111–113, 117
Carbon capture, 173, 240
  cost of, 114
  and storage, 115, 217–218
Carbon Clock, 215–216
Carbon cycle, 26–30, 28f
  big numbers, visual representation of, 28–30, 29f
  fast cycle, 27
  gigametric ton, 29f
  slow cycle, 27
Carbon dating, 69
Carbon dioxide (CO_2), 3–4, 7–8, 64, 83–84, 114, 270
  in atmosphere
    preindustrial age mass of, 275
    total mass, 275
  capture systems development, 212
  concentration, 24f, 25, 41–42, 64, 99, 174–175
  convert into synthetic fuels, 212
  natural sources of, 72
  per capita, 36f
  into usable liquid fuels, 137–138
Carbon dioxide (CO_2) emissions, 14, 36, 37f, 114, 116, 126, 188, 204–209, 205f, 217–218
  global, from fossil fuels, 71f
  rising, 40
Carbon dioxide equivalent, 64, 151t, 225t, 289, 296–297
Carbon emissions fee, 16
Carbon Engineering, 138
Carbon fee, 211–212, 223–226, 237, 241, 244, 253, 291
  for United States, 225
  $US50/ton of CO_2 carbon fee on fuels, 225t
Carbon-Fee Equivalent Rebates, 297
Carbon-Fee-Equivalent Tariffs, 297
Carbon Fees Trust Fund, 291, 296
Carbon footprint, 73
Carbon-free energy, 128–129, 191
Carbon-free source of liquid fuels, 172
Carbon monoxide, 72
Carbon removal, 115

Carbon sink, 14, 111, 134, 170–171, 244–245
Carbon tax, 125, 192, 211–212, 223–226, 225t
Carbonic acid, 45, 66, 83, 276
Carteret Islands, 82
Cash for clunkers program, 249
Caterpillar, 303
Cattle raising, 65t
Cement, manufacturing, 207–208
Center for Climate Integrity, 82
Central Valley, California, 155–156, 197
Changing political landscape, be aware of, 252
Charcoal, 135
Chemical weathering, 67
Chernobyl, 189
Chersky, 86
Chevron Corporation, 303
Children's lawsuits, litigation, 298
China, 156–158, 197–198, 217, 233–234, 246–248
  automobile industry, 157–158
  coal, 157
  energy security, 157
  greenhouse gas emissions, 156
  solar market, 158
  Tengger Desert plant, 158
Chlorofluorocarbons, 185–186
Chu, Steven, 11–12
Chubb, 251–252
Cigarette smoking and cancer, 186–187
Cigarette tax, 186–187
Citizens' Climate Lobby (CCL), 250–251, 296
  border adjustments, 297
  collection of carbon fees/Carbon Fees Trust Fund, 296
  emissions reduction targets, 296–297
  equal per-person monthly dividend payments, 297
City and county lawsuits, litigation, 298
Civil unrest, 95
Clean Air Act of 1990, 184, 195
Clean Power Alliance, 147
Climate accountability initiative, 297
Climate change, 11–12, 89, 242–243

Climate change (*Continued*)
activists, engage with, 250–251
deniers, 175–176
migrations caused by, 96–97
refugees, 81
skepticism, 174–176
vs. weather, 76–78
Climate model, 97
Climate research, 99
The Climate Science Special Report, 109–110, 242–243
Cloud cover, 129–130
Coal, 16, 27, 125–126, 128*t*, 152–153
Coal burning power plant, Iowa, 3*f*
Coal-fired fossil fuel plants, 190
Coal-fired generating capacity, 140
Coal-fired power plants, 3, 143, 157, 159–160, 173, 207
Coastal flooding, 43–44, 82, 90, 121–122, 286
Coconut oil, 135
CO_{2eq} gas components, 66*f*
Coffin nails, 186
Cold War, 259–260
Columbia Glacier, Alaska, 47*f*
Combustion engine, 8, 184
Combustion process, 126
Commercial buildings, energy use, 3, 207–208
Commercial nuclear power, 6–7, 129, 189, 259–260
Community Choice Aggregators (CCAs), 147
Competitors to traditional utilities, 147
Complete decarbonization, 110–111
Compressed air energy storage (CAES), 140
Concentration of CO_2, 24*f*, 25, 41–42, 64, 99, 174–175
Concerned citizens, actions for
changing political landscape, be aware of, 252
divest from major carbon emitters, 251–252
educate yourself and others, 250
engage with climate change activists, 250–251

individual actions, 252
litigation, 251
lobby for government action, 251
Congress, U.S., 182–183, 296–297
Congressional committee, 245–246
Conversion factors, 271–272
Cooling water, 118, 189, 192–193, 204–209
Copenhagen, 101
Copenhagen Accord, 104
Coral reefs, 27
Corn-based ethanol, 134
Corn crop, ethanol production, 195
Corning Inc, 305
Corporate annual reports
Alliant energy, 303–304
Bank of America, 301
Caterpillar, 303
Chevron Corporation, 303
Corning Inc, 305
DTE Energy Inc, 304
Duke Power, 305
DuPont Corporation, 302
Eaton Corporation, 305
Eli Lilly Corporation, 301
Home Depot, 303
Olin Corporation, 301
Southern Company, 302
UPS Corporation, 304
Verizon, 302
Wisconsin Electric (WEC Energy Group), 304
Corporate naysayers, 255–256
Corporate philosophies, 254
Cosmic rays, 69
Cost competitive, 124, 136*t*, 192
Cost differential, 16, 223–224, 249–250
Costa Rica, 58
Cost-effective, 129–130, 172, 175, 248–249
Crop failure, 120–121, 219, 294–296
Crop yields, 75, 120, 174–175, 203–204, 219
Cropland, 76, 95
Cuban Missile Crisis, 259

Index   311

D
Dangers of global warming, 5—6
DARPA. *See* Defense Advanced Research
    Projects Agency (DARPA)
Decarbonization, 113
Decommissioning process, 193
Defense Advanced Research Projects
    Agency (DARPA), 202, 248
Deficit, 173—174, 244
Deforestation, 64, 209, 210f, 244—245,
    276
Del Mar, California, 220, 286—287
Demand, electrical, 143—144
Department of Defense, 95, 242—243
Department of Energy, 153, 245—246,
    291, 296
Depletion allowance acts, 229
Deregulation, 189—190
Desertification, 87
Deserts and tropics expand, 55—57, 87
Design life, 193, 236
Destruction, 3, 111, 209—212, 221—222
Developing countries, energy requirement
    by, 37, 37f
Deviation in sea ice extent, 52f
Diesel fuel, 3, 72, 115, 137, 150, 187, 195,
    223—224
Diesel-powered vehicles, Europe,
    187—188
Digital technology, 143—144
Direct-air capture and storage, 115
Direct current (DC) distribution system,
    142
Diseases, spread by global warming,
    92—93
Distribution system, 142—145, 180,
    231—232
Droughts, 90—92, 120
Dry storage, 194
DTE Energy Inc, 304
Duke Power, 305
DuPont Corporation, 302

E
E3 Carbon Tax Calculator, 225
Early warning signs, global warming,
    58—60

Earth
    carbon cycle, 26—30, 28f
        big numbers, visual representation of,
            28—30, 29f
        fast cycle, 27
        gigametric ton, 29f
        slow cycle, 27
        greenhouse effect, 25—26
        incident solar radiation, 21—23
        Milankovitch cycles, 23—25, 24f
        temperature increase, 30, 30f
Earth's average temperature rising, 40—41,
    78
Earth's cycles, 26—27
Earth's temperature, 2, 4, 78, 107, 174, 239
Earth's total energy use, 231—232, 234
East Asia, 79
East Germany, 149—150
Eaton Corporation, 305
Economy of scale, 172, 194, 227, 246—247
Efficiency, energy, 16, 36—37, 94, 204,
    254—255, 305
Eisenhower administration, 189
Electric energy, 16
Electric grid, 142—143
    challenges, 143
    components, 142—143
    one-way grid, 143
    time-of-use, 143
    two-way flow of power, 143
Electricity pricing models, 191
Electricity storage, 138—139
Electric transmission and distribution
    systems, 172
Electric vehicles, 118, 208, 248
Electrolysis, 137—138
Elevated CO_2, 92
Eli Lilly Corporation, 301
Eliminate, 7—8, 14, 170—171, 216, 240
Emission control, 184, 188
Emissions Gap Report, 102—103
Emissions reduction targets, CCL,
    296—297
Energiewende, 149
Energy alternatives
    advantages of, 123
    disadvantages of, 123

Energy alternatives (*Continued*)
 fossil fuels, 129–147
  coal, 125–126
  natural gas, 127–128
  oil and its derivatives, 126–127, 126*f*
  types, 128*t*
 global energy consumption
  by fuel, 123*f*
  by sector, 124*f*
 nuclear power, 128–129
 renewable energy, 129–147
  biofuels, 133–134
  CO_2 into usable liquid fuels,
   137–138
  competitors to traditional utilities, 147
  electricity, advantages of, 130
  energy storage, 138
  hydroelectricity, 132
  hydrogen, 137
  LCOE utility-scale PV solar, 132*f*
  LCOE wind power, 132*f*
  micro grid, 145–147
  primary sources of, 136, 136*t*
  smart grids, 142–143
  solar energy, 130
  wind power, 131, 133*f*
Energy and GHG equivalencies, 272–273
Energy carrier, 137
Energy efficiency, 16
 and conservation, 204, 206*f*
Energy forecast, 224–225
The Energy Independence and Security
 Act of 2007, 195
Energy Innovation and Carbon Dividend Act
 (HR763), 244, 290
Energy plus the Environmental Protection
 Agency, 242–243
Energy Policy Act of 2005, 195
Energy Reliability Council of Texas
 (ERCOT), 180
Energy security, 158–159
Energy storage, 138, 140–141, 172
 applications for, 138
 batteries, 141–142
 compressed air energy storage (CAES),
  140
 lithium-ion battery, 139

 pumped hydro, 141–142
Energy-saving method, 253–254
Energy transformation, 149
Energy turnaround, 149
Engine performance, 135–136, 184
England, 2, 26, 39, 89
Enhanced oil recovery, 115
Enterprise, 189
Environmental Health Hazard Assessment,
 58–59
Environmental Protection Agency, 184,
 245–246
Equal per-person monthly dividend
 payments, CCL, 297
Ethanol, 129, 134, 195–197, 196*f*
Ethiopia, 35*t*, 210*f*
Europe's push for diesel vehicles, 187–188
European Union, 71, 135, 149, 232, 241,
 244–245
Extinction Rebellion (XR), 294–296
Extraction, 85, 126–127
Extreme floods, 120
Extreme weather events, 54–55
ExxonMobil, 298

F

Fast carbon cycle, 27
Fee and dividend, 290–291
Feed-in tariff (FIT), 150–151
Feedstock, 62, 127, 128*t*, 133–134
Fertilizer use, 65*t*
Field studies, 58, 96
Financial crisis, global warming, 94
Financial measures
 Cap and Trade program, 289–290
 fee and dividend, 290–291
First-generation biofuels, 134
Fisheries, 83–84
Flood and sea rise mitigation, 285–287
 managed retreat plan, 286
Flooding, sea levels rise, 79–83, 81*f*
Florence, 90
Florida, 14, 79–80, 220, 287, 299
Flow battery, 141
Fluorinated gases, 3
Fluorocarbons, 170–171
Flywheel, 142

Index   **313**

Forests, fires, logging, 120–121
Fossil fuels, 10, 14, 16–17, 25, 31, 34–37,
    62, 70, 118, 129–147, 160–161,
    164, 169, 204–209, 227, 230
  coal, 125–126
  energy content, 135t
  global carbon dioxide emissions from,
    71f
  global economy, 12–13
  hard to replace, 170–171
  heavily subsidized, 164, 237
  increases in atmospheric CO_2 correlates
    with, 72
  natural gas, 127–128
  oil and its derivatives, 126–127, 126f
  subsidies, 236t
  types, 128t
450-MW High Lonesome project, 180
Fourth National Climate Assessment, 10
Fracking, 126–127, 141, 154, 179,
    223–224, 228–229
France, 190
French riots, 217
Fresno, 155–156
Friday's for Future (FFF), 293–294
Fuel cost, 138, 159, 228, 233–234,
    253–254
Fuel economy standard, 182–183
Fuel efficiency, 208, 254–255
Fuel injection, 182–183
Fukushima nuclear power station, Japan,
    152, 160

G

Gallon of ethanol, 134
Gas-fired fossil fuel plants, 190
Gasoline, 134, 137, 195–196, 224, 244,
    253, 272–273
Gas turbine, 130, 141, 143
Gasoline, 3, 62, 115, 157, 195, 217, 253
Geoengineering, 218–219
Georgia, 191–192
Germany
  average cost of electricity in, 151
  coal-fired power plants, 151–152
  Energiewende program, 149, 152
  Federal Administrative Court, 188

  in greenhouse gas emissions, 149–150,
    150f
  greenhouse gas reductions, 151t
Gigajoule (GJ), 21, 21b
Gigametric ton, 28, 29f
  of carbon, 28b
Glacier National Park, 53, 58
Glaciers, melting of, 47–54, 59, 84
Global atmospheric temperature rate of
    change, 41f
*Global Covenant of Mayors for Climate and
    Energy*, 102
Global economy, 241
  powered by fossil fuels, 12–13
Global emissions of CO_{2eq} in 2018, 108f
Global energy consumption
  by fuel, 123f
  by sector, 124f
Global energy efficiency, 118
Global inequities, 164
Global warming, 9–10, 14, 61, 75–76,
    82b, 86b, 117, 161, 239, 259–260
  agriculture, droughts, loss of cropland
    and wildfires, 90–92
  air pollution increase, 79
  Arctic, 45–46
  atmospheric CO_2 concentration, increase
    in, 41–42
  China, 156–158
    automobile industry, 157–158
    coal, 157
    energy security, 157
    greenhouse gas emissions, 156
    solar market, 158
  climate change
    migrations caused by, 96–97
    skepticism, 174–176
    vs. weather, 76–78
  CO_2 emissions rising, 40
  deserts and tropics expanding, 55–57,
    87
  deviation in sea ice extent,
    52f
  early warning signs, 58–60
  earth's average temperature rising,
    40–41, 78
  educating the public, 164–165

Global warming (*Continued*)
extreme weather events, increasing, 54–55
financial crisis, 94
fossil fuels, 164
hard to replace, 170–171
heavily subsidized, 164
frequency and severity of storms, 89–90
fundamental forces
absolute and per capita emissions, 34–35, 35*f*, 35*t*
energy requirement, by developing countries, 37, 37*f*
global population rising, 31, 32*f*
gross domestic product growth and energy, 36–37, 36*f*
inequities, United Kingdom and United States, 31–32
living standards, 31, 37
pivotal position of U.S., 33, 33*f*
Germany
average cost of electricity in, 151
coal-fired power plants, 151–152
Energiewende program, 149, 152
in greenhouse gas emissions, 149–150, 150*f*
greenhouse gas reductions, 151*t*
glaciers, melting of, 47–54, 84
global economy, by fossil fuels, 12–13
health problems, severe, 92–94, 93*f*
ice caps, melting of, 47–54, 84
India, large population, little energy, 158–159
coal-fired power plants for electricity production, 159
per capita energy, 158
international cooperation, 163–164
Japan, 159–160
latency, 11–12, 13*f*, 76
media, 165–167
migration, rising temperatures, 58
national security implications, 95
ocean acidification, 45
oceans become more acidic, 83–84
ocean temperatures rise, 42–43, 43*f*
permafrost melts, 54, 85–86
plant, animal, and human migration, 58

political leaders, 176
positive message, 169
possibility of failure, 176–177
public support for government action, 169–170
public uncertainty, 167–169
Russia, 160–161
sea ice melting, 47–54, 47*f*, 84
sea levels rise, 43–44, 44*f*
flooding, 79–83, 81*f*
solving technical challenges, 171–173
carbon capture, 173
carbon-free source of liquid fuels, 172
electric transmission and distribution systems, 172
intermittent nature of renewable electricity from solar and wind, 172
nuclear power, 173
species migration and extinction, 87–89, 88*f*
strong and growing economies, 173–174
subsidence, 85–86
temperature increases due to, 119–120, 119*f*
tipping points, unanticipated changes, 97–98
top priorities, 241–242
unachievable international cooperation, 13
United States fails, leadership position, 153–156
U.S. participation, 14
Goodell, Jeff, 83, 286–287
Government actions, 242–250
cash for clunkers program, 249
educating the public, 242–243
fee for carbon emissions, 244
join international effort to combat global warming, 244–245
large-scale demonstration projects, 248
let failures die, 249
mitigation, 250
national plan, 245–246
new laws and regulations, 247
report card, 243–244
siting, 247

Index

subsidies, 248
technology plan, 246—247
Great Barrier Reef, 83
The Great Depression, 259
Great Eastern Japan Earthquake, 160
Green agricultural practices, 211
Green energy, 137, 155, 180, 240, 302
Greenhouse gas, 10—11, 64, 125, 276
  equivalencies, 273
  sources, 65*t*
Greenhouse gas effect, 10—11, 25—26, 39,
  61—62, 165
Greenhouse gas (GHG) emissions, 9—10,
  14, 16, 100, 170—171, 190,
  239—240, 244—245, 257, 289, 291
  China, 156
  forecast of, 120—121
  Germany, 149—150, 150*f*
  IPCC report, 100—101
  Paris Agreement, 101—102
  in U.S., by economic sector, 34*f*
Greenland, 40, 45, 47, 49*f*, 54, 84, 98
Gross domestic product (GDP), 154, 171,
  203—204, 205*f*, 216, 235
  growth and energy, 36—37, 36*f*
Ground water extraction, 85
Gulf Coast, 85
Gulf of Mexico, 77, 82, 90
Gulf Stream, 42, 45, 98

H

H.R. 763, Energy Innovation and Carbon
  Dividend Act of 2019, 244, 290,
  297
Habitat, 66, 84, 89
Half life, 69
Hansen, James, 114, 165, 239, 276,
  298—299
Hardin, Garrett, 163—164
Hawaii, 4, 40, 86, 102, 145, 155, 248
Health problems, 73, 92—94
Heat, 2, 12, 21—22, 43, 62, 118, 125,
  207—208, 230
Heat stress, 78
Heat stroke, 78
Heat wave, 12, 54—55, 75, 78, 119—120
High cost of doing nothing, 121—122

High overshoot scenario, 112—113
High-speed rail, 197—198
Hindu Kush-Himalayan region, 52
Hole in ozone layer, 185—186
Holland, 39
Home Depot, 303
Hong Kong, 122
Hoover Dam, 132—133
Housing and Urban Development (HUD),
  219
Houston, 77, 90, 222
Hubbard Glacier, Alaska, 85*f*
Hubble telescope, 259—260
100-year storm, 12, 76, 90, 222
Human waste, 90
Humidity, 121, 174—175
Hunter-gatherer, 73
Hurricane Harvey, 77, 90, 95
Hurricanes, 54, 76, 89—90
Hybrit feasibility study, 207—208
Hydro, 124
  advantages, 136*t*
  disadvantages, 136*t*
Hydrocarbons, 126
Hydroelectricity, 62, 132
Hydrogen, 137, 141, 212, 240, 272—273
  advantages, 136*t*
  disadvantages, 136*t*
Hydrogen bomb, 259
Hydrogen-powered, 17, 137, 150
Hygiene, 92—93
Hypoxia, 83—84

I

Ice age, 23, 39, 99—100
Iceberg, 48
Ice caps, melting of, 47—54, 84
Ice core measuring techniques, 23—25
Iceland, 51, 248
Ice melt, 48—49, 77
Ice sheet, 43, 48
Ice shelf collapse, 48
Incentives, to reduce carbon, 223—224
India, large population, little energy,
  158—159
  coal-fired power plants for electricity
  production, 159

India, large population, little energy (*Continued*)
  per capita energy, 158
Indirect energy, 73
Industrial processes, 3, 16, 170–171, 207–208
Industrial Revolution, 2, 26, 31, 40, 69–71, 275–277
Industry, actions for
  changing corporate philosophies, 254
  proactive about fighting global warming, companies, 254–255
  corporate naysayers, 255–256
  keep innovating, 253–254
  small group of bad actors, 256
Infectious diseases, 89
Influx, 97
Infrared radiation, 22, 39
Infrastructure project, 173, 180, 244, 250
Insects, and global warming, 92–93, 96
Insulation, 11, 207–208, 223
Insurance, nuclear power, 194
Intergovernmental Panel on Climate Change (IPCC), 9, 12–13, 30, 75, 100–103, 163, 175–176, 201, 214–216, 240, 276, 296–297
  alternative scenarios, 107–110, 109*f*
  baseline, 30
  Fifth Assessment, 43–44
  history of, 103–104
  overshoot and return to temperature target, 113*f*
  450 ppm, CO_2 concentration of, 277–278
  scenarios, 111–114, 112*t*, 113*f*
  special reports, 102–103, 215
Intergovernmental Science Policy Platform on Biodiversity and Ecosystems Services (IPBES), 87
Internal combustion engine, 8, 17, 118, 184
International cooperation, 163–164
International efforts to global warming
  early efforts, 99–100
  Intergovernmental Panel on Climate Change (IPCC), 100–103
  history of, 103–104

  special reports, 102–103
  Paris Agreement, 101–102
International Energy Agency (IEA), 235
International Global Atmospheric Research Program, 99
International Monetary Fund (IMF), 235–236
International regulation, 185
International system of units (SI), 270–271
  advantages, 270–271
  units for energy and power, 270
International units, 21*b*
Iranian Revolution, 181–182
Islanding, 146–147
Isle De Jean Charles, 219
Isotope, 68*b*
Italian Alps, 53
Italy, 51, 53, 145

J

Japan, 159–160
Jatropha oil, 135
John Kerry's Initiative, 297
Join international effort to combat global warming, 244–245
Joule (J), 21*b*, 270
Juliana versus United States, 298–299

K

Katrina (hurricane), 89
Keeling, Charles, 4
Kelly, Walt, 72
Kennedy, President, 201
  assassination, 259
Kerosene, 272–273
Kolyma River, 86
Korea, 125
Korean War, 259
Kyoto Protocol of 1997, 100, 104, 187

L

Lake Tahoe, California, 59
Land clearing, 64
Landfill, 135
Land-use changes, 3, 107
  and agriculture improvement, 211
Land use practices, 17, 87, 244–245

Index 317

Large-scale demonstration projects, 248
Larson C Ice Shelf, 48
Latency, 1, 11−12, 13*f*, 65, 76, 176−177
Leaded gasoline, 184
Legislation and regulations, 260
Less developed countries, 92, 135
Levelized cost of energy (LCOE),
191−192, 228
alternatives in North America, 229*f*
utility-scale PV solar, 132*f*
wind power, 132*f*
Libby, Willard, 69
Life cycle costs, 228
Lifestyle choices, 73
Light bulbs, 11, 167−168, 170
Liquid fuels, 208
Liquid hydrocarbon fuel, 138
Liquified natural gas (LNG), 157, 159
Lithium-ion, 140
Lithium-ion battery, 139, 146−147
Litigation, 251, 298−299
children's lawsuits, 298
city and county lawsuits, 298
state lawsuits, 298
Living standards, 34−35, 79, 156,
203−204, 208, 211, 240
Load shedding, 144
Lobbying, 197
Lobby for government action, 251
Long-wavelength infrared radiation, 22
Los Angeles, California, 176, 184, 193, 197
Loss of cropland, 90−92
Louisiana, 85, 219
Low income household, 290
Low or no overshoot scenarios, 112−113
Low power density, 230
Low-lying areas, 43−44, 80−81, 222, 286
Low-lying island, 43−44
Lung cancer, 186
Lyme disease, 89

M

Maine, 79−80, 248
Major carbon emitters, divest from,
251−252
Managed retreat plan, 286
Manhattan Project, 260

Man-made CO_2, 27, 68*f*
atmospheric CO_2 correlates with fossil
fuel, increases in, 72
carbon-14 (C-14), 68−70
history, 69
Libby's discovery, 69
carbon footprint, 73
clearing tropics, 65*f*
CO_{2eq} gas components, 66*f*
deforestation, 64
fossil fuel, 62
global CO_2 emissions by world region,
67*f*
greenhouse gas sources, 65*t*
Industrial Revolution, 70−71
Man-made global warming, 110
Manual labor, effect of global warming, 78
Marine life, 66
Marine organisms, 83−84
McKibben, Bill, 61−62, 251−252
Media, 165−167
Mediterranean Sea, 96−97
Mental health and global warming, 92
Melomys rubicola, 89
*Mercator Research Institute on Global
Commons and Climate Change*
(MCC), 215−216
Merced, California, 155−156
Meteor, 87
Methane (CH_4), 3, 46, 65*t*, 78, 86,
121−122, 127, 240, 272−273
Methane clathrate, 77
Methyl tertiary-butyl ether (MBTE), 195
Mexico, 48, 57, 77, 82, 90, 179, 291
Miami, Florida, 287
Michigan legislation, 304
Micro grid, 145−147, 145*f*
Microorganism, 77
Middle East, 87, 152−153, 181, 254
Migrate, 6, 121
Migrations
climate change, 96−97
rising temperatures, 58
Milankovitch cycles, 23−25, 24*f*
Mileage, of vehicles, 208
Mitigation, 219−223, 221*f*
flood and sea rise, 285−287

Mitigation (*Continued*)
government actions, 250
managed retreat plan, 286
measures, 220–222
sea wall construction, United States, 220
Molten salt reactor, 142
Montana, 58
Monteverde National Park, 58
Montreal Protocol, 185–186
Montreal Protocol of Vienna Convention, 100
Moon shot, 201–202
Moscow, 160
Munger, Charlie, 223–224

N

National Aeronautics and Space Act, 202
National Aeronautics and Space Administration, 242–243
National debt, 174, 244
National Defense Education Act, 202
Nationally Determined Contributions (NDCs), 102–103
greenhouse gas reduction pledge, 110
National Oceanic and Atmospheric Administration (NOAA), 30, 42
National plan, 245–246
National security, 97, 202
National security implications, 95
Natural disasters, 95
Natural gas, 127–128, 128t, 151, 154–155, 159–160, 207, 223–224, 233–234, 272–273
Natural gas leakage, 65t
Natural gas reforming, 137
Natural gas-fired power plants, 3
Nepal, 51
Net zero emissions, 109, 201, 219
Net zero, defined, 110–111, 111b
Network, 130, 197–198, 204
Nevada, 230, 248
New Delhi, 12
New England, 247
Newport Beach, California, 82, 285
NextEra Energy Resources, 254
Next-generation biofuels, 135–136, 197

Newsom, Gavin, California Governor, 155–156
Nigeria, 35t
1970s oil price hikes, 181–183
1973 oil embargo, 204–209, 253–254, 291
Nitrogen fertilizer, 211
Nitrous oxide (N_2O), 65t
Nobre, Carlos, 120–121
Nonlinear climate effects, 77
Nonradioactive waste, 194
Nonviolent civil disobedience, 294–296
Norfolk, Virginia, 80, 95
Norilsk, 86
North Pole, 46–47
North Sea, 182, 254
Northern Hemisphere, 45–46, 54, 79, 277
Northern latitude, 5
Northwest Passage, 46, 48–49
Norway, 46, 47f, 248
Nuclear isotopes, 192
Nuclear power, 128–129, 173, 259–260
Action plan, 212
decommissioning process, 193
financing, 194
future for, 191–195
insurance, 194
nuclear power plants, 191–195
recycling spent nuclear fuel, 192
spent reactor fuel, 194
in United States, 188–191
water-cooled condensers, 192–193
Nuclear power plants, 129, 143, 158, 161, 173, 189–195, 236
Nuclear Regulatory Commission safety requirements, 190, 194–195
Nuclear weapons, 69, 189, 259–260
Nutrition, 93–94

O

OAPEC, 181–182
Ocean acidification, 45
Ocean temperatures rise, 42–43, 43f
Oceans, acidic, 83–84
Offsetting benefit, 240
Ohio, 167
Oil, 128t, 166

and its derivatives, 126–127, 126f
production, 153
Oil-based fuel, 208
Oil-based products, 3
Oil drilling technique, 141
Oil extraction, 85
Oil price hikes, 181–183
Oil-producing countries, 182, 229
Olin Corporation, 301
One-child policy, 37
Organization of the Petroleum Exporting Countries (OPEC), 181
Our Children's Trust, 298–299
Over-grazing, 55–57, 163–164
Overshoot pathways, IPCC scenarios, 112
Oxygen depletion, 83–84
Ozone depletion, 185
Ozone layer, 185
Ozone layer, hole in, 185–186
Ozone-destroying gases, 100

P

Pacific Gas and Electric (PG&E) transmission line, 146
Pacific Ocean, 4, 46, 80–81
Palm oil, 209
Panama Canal, 260
Panic-driven decarbonization, 94
Paraguay, 248
Paris Agreement, 14, 101–104, 108–109, 113, 116–117, 153–154, 156–157, 160, 163, 204, 244–245, 251, 293–294, 297
Peaking unit, 143
Peer review, 175
Penguins, 84
Pentagon, 259
Per capita emissions, 34–35, 35f, 35t
Per capita energy, 164, 203–204
Perfluorocarbons (PFCs), 33, 291
Permafrost, 5, 54, 77
Peru, 51
Permafrost melting, 46, 54, 85–86
Permian Basin, 179–181
Petroleum-based transportation fuels, 208
Pew Research, 167
pH, 45

Photosynthesis, 21–22, 65–66, 276
Phytoplankton, 66
Pipelines, 127–128, 204–209, 235, 237, 247
Plant life, 9, 64, 134, 209
Pledges under Paris Accord. *See* Nationally Determined Contributions (NDCs)
Polar bear, 47f, 84
Polar ice, 5, 40, 77, 84, 112, 176–177
Polio vaccine, 259–260
Political leaders, 167, 297
Pollution, 157
Population growth and global energy demand, 32f
Portland General Electric, 141
Positive feedback, 98
Potential of Hydrogen (pH), 45
Power of 10, 28–30
Power purchase agreement, 102
Power the earth, 232
with solar energy, 231
Pre-Industrial Revolution, 96
Precession, 23
Price-Anderson Nuclear Industries Indemnity Act, 194
Private sector economy, 173
Process energy, 224
Protons, 68
Protozoa, 90
Public education, 164–165, 242–243, 250
Public health, 17, 93–94, 175–176, 187, 251, 289–290
Public investment, 248
Public policy, 168f
Public support, 7, 164–165, 169–170, 227, 257
Public transportation, 167–168
Public trust doctrine, 299
Public uncertainty, 167–169
Public utility commission, 249–250
Pumped hydro, 140–142

R

Radiant energy, 2
Radioactive spent fuel, 190
Radioactive waste, 114, 124, 189–190, 192, 194

320 Index

Railroads, 208
Rainfall, 11, 45–46, 50, 77, 87, 219, 222
Ranching, 120–121
Rapeseed, 135
RE100, 254–255
Reabsorbed, 133–134
Real energy prices, 183
Rebate, 290–291, 297
Recession, 181, 203–204, 260
Recycle, 115, 137, 161, 192, 248
Refinery, 289
Reforestation, 209–212
Refrigerants, 185
Regulatory agency, 155
Reliability, 144–145, 189
Renewable energy, 16, 71, 73, 102,
 129–147, 151–153, 155, 203, 227,
 241, 289
 biofuels, 133–134
 China, 157–158
 CO_2 convert into synthetic fuels, 212
 CO_2 into usable liquid fuels, 137–138
 competitors to traditional utilities, 147
 cost of, 249
 electricity, advantages of, 130
 energy storage, 138
 hydroelectricity, 132
 hydrogen, 137
 LCOE utility-scale PV solar, 132f
 LCOE wind power, 132f
 micro grid, 145–147
 power the world, 230–232, 231t
 primary sources of, 136, 136t
 smart grids, 142–143
 solar energy, 130
 wind power, 131, 133f
Renewable Energy 100% (RE100),
 254–255
Renewable Fuel Standard, 195
Report card, 243–244
Residential buildings, 3, 207–208
Resilience, 144
Rice farming, 65t
Rink Glacier, 53–54
Rising sea levels, 6, 12, 83, 104, 174, 219,
 250
Rodents, 92–93

Roman, 39
Ross Sea Ice Shelf, 48
Russia, 160–161
Russian Federation, 35t

S

Sahara desert, 230–231
Sahel desert, 55–57
Saltwater intrusion, 43–44, 80–81
San Diego, 59, 82, 146, 220
San Francisco, 176, 197, 298
San Francisco Bay, 59
San Juan Capistrano, 82
Santa Barbara, 59, 82
San Onofre Nuclear Generating Station,
 193
Saturation levels, 83–84
Saudi Arabia, 14, 126–127, 152–153, 179
Scripps Institution of Oceanography, 4
Sea ice melting, 47–54, 47f, 84
Sea levels rise, 43–44, 44f, 120, 174
 flooding, 79–83, 81f
Sea surface temperature, 42, 43f
Sea wall, 220
Secondhand smoke, 186
Service life, 129
Seven Sisters, 181
Sewage treatment, 73, 135
Shah (of Iran), 182
Shanghai, 80, 220
Shedding, 144
Shellfish, 6, 66, 83–84
Shismaret, 286
Siberia, 45–46, 86, 91, 209
Sierra Nevada Mountain range, 155
Signs of global warming, 5
Sink, for carbon, 14, 111, 134, 244–245
Siting, government actions, 247
Sixth Assessment Report, 103
Slow carbon cycle, 27
Smart electrical grids, 191
Smart grids, 142–144
 benefits, 144
 Borrego Springs micro grid, 145f
 development, 144
 with smart meters, 144
Smart meters, 144

Index

Smoke free zone, 186
Snowpack, 120
Solar and wind farms, North Palm Springs, California, 228*f*
Solar and wind power costs, 233*t*
Solar arrays, 247
  with battery storage, 180
Solar cells, 10, 16, 131
Solar energy, 21–22, 129–130, 149–150, 191, 230
  advantages, 136*t*
  disadvantages, 136*t*
  for electricity production in United States, 154
  sources and sinks, 22*f*
Solar farm, 141, 179, 231, 247, 254
Solar panel, 141, 227, 230–231
Solar photovoltaic energy production, 131
Solar power, 143
  supply and demand, 131*f*
Solar radiation, 2
  incident, 21–23
Solar Star power plant, 139–140
Solomon Islands, 80–82
South America, 48, 209
Southern Company, 302
South Pacific islands, 220
Soviet Union, 202
Soybean, 120–121, 135
Space program, 201–202
Space race, 259
Spain, 145
Species migration and extinction, 87–89, 88*f*
Spent fuel, 190, 192, 194, 212
Sports utility vehicles (SUVs), 183
Sputnik, 202, 259
SR1.5, 109–110, 112
Standard of living, 9, 34, 125, 158, 169
Stanford University, 230
State Air Resources Board, California, 154–155
State lawsuits, litigation, 298
Stein, Herb, 117
Storms and flooding mitigation, 222
Storms, frequency and severity of, 89–90
Storm surge, 54

Stranded asset, 7, 94, 159, 251–252
Student Climate Network, 293–294
Submarine, nuclear power, 129, 189
Subsidence, 85–86
Subsidy, 16, 125, 164, 191, 197, 224, 236*t*
Subsistence farmer, 75–76
Suess, Hans, 69
Sugar cane, 133–134, 195–196
Sulfur dioxide, 218–219
Sunlight, 39
Sunset Rule, 197
Surplus power, 130
Swamp gas emissions, 65*t*
Swedish steel plant, 207–208
Switzerland, 51, 53, 293–294
Swords into Plowshares project, 189
Synthetic fuels, 172, 240
Synthetic liquid fuel, 115

T

Tankers, 204–209, 237, 247
Technology plan, 246–247
Temperature anomaly, 26, 27*f*
Temperature increase, 30, 30*f*
Temperature zones, 6, 57–58
Tengger Desert, 158
Terrorism, 95, 245–246
Texas, 77, 115, 155, 179–180, 226, 247–250
Thermodynamic inefficiencies, 16
Threat multiplier, 95
Three Mile Island, 189
Tidal flooding, 80
Time-of-use, 143
Tipping point, 6, 97–98, 107–108, 239–240, 259
Tobacco, 186
The Tobacco Master Settlement Agreement, 187
Toshiba, 189
Total CO_{2eq} emissions, 34
Total global greenhouse gas emissions, 107
Tragedy of the Commons, 117–118, 163–164
Transcontinental railroad, 260
Transmission system, 142–145, 150–151
Transportation system, 241
Trapped methane, 5, 54

"Trend is our friend", 227—229
Trump administration, 101—102, 108—109, 297, 299

U

Ultraviolet (UV) radiation, 185
Underdeveloped countries, 73
Underground storage, 8, 114—115, 170—171
Unemployment, 171
United Kingdom, 31—32, 91, 140—141, 294—296
United Nations Environmental Program (UNEP), 55, 102—103
United Nations Framework Convention on Climate Change (UNFCCC), 100, 109—110
United Nations General Assembly, 186
United Nations Organization, 202
United States, 174
   estimated cost of seawall construction in, 221f
   ethanol producer, 195—196
   media, 166
   participation in global warming, 14
United States Climate Alliance, 102
United States fails, leadership position, 153—156
Units and conversion factors, 28—30, 270—272
Unqualified naysayers, 166
UPS Corporation, 304
Urban air pollution, 128t
U.S. Climate Science Special Report, 10, 61, 110, 242—243
U.S. Department of Defense, 95
U.S. Department of Energy, 67—68, 153, 296—297
U.S. Department of Housing and Urban Development, 219
U.S. electricity storage capacity, 140
U.S. Energy Information Agency, 191—192
U.S. Environmental Protection Agency, 73
U.S. Food and Drug Administration, 186—187

U.S. Global Climate Change Research Program, 242—243
U.S. government, 10, 95, 110, 153—154, 189—190, 192, 243—244, 251
U.S. greenhouse gas emissions, 33f
   by economic sector, 34f
U.S. National Academy of Science, 99, 185
U.S. National Aeronautics and Space Administration (NASA), 242—243
U.S. National Defense Education Act, 202
U.S. National Oceanic and Atmospheric Administration (NOAA), 79—80, 99
U.S. Navy, 80, 95, 129
   nuclear power program, 189
U.S. public policy priorities 2019, 168f

V

V.C. Summer nuclear plant, 191
Vegetable oil, 135
Vehicle emissions, 157
Vehicle emissions standards, 154—155
Verizon, 302
Vienna Convention for the Protection of the Ozone Layer, 185
Vogtle nuclear power plant, 191—192
Volkswagen, 188
Voting population, 170

W

Waste decomposition, 65t
Water-borne diseases, 90
Water-borne infectious diseases, 83
Water cooling, 192—193
Water supplies, 17, 59, 83, 120, 157
Water vapor, 54—55, 126, 138
Watt (W), 270
Wave damage, 82
Wave length, 22, 25—26, 39
Weather $vs.$ climate change, 76—78
West Germany, 149—150
Westinghouse, George, 142, 189
Wetlands, 55—57
Wet storage, 194
Wildfires, 59, 90—92
Wind energy, 129—131, 149—150, 191

Index 323

advantages, 136*t*
disadvantages, 136*t*
for electricity production in United
States, 154
Wind farm, 131, 179, 228*f*, 232–233, 247
Wind generating farm, Desert Hot Springs,
California, 15*f*
Wind power, 131, 133*f*, 143, 232–233
Wind speed, 232
Wind turbine, 10, 124, 152, 227, 232, 249
Wisconsin Electric (WEC Energy Group),
304
Wood pellets, 133–134, 152

World Bank, 75–76, 156, 235–236
World Climate Research Program, 99
World Climate Strike, 293–294
World Energy Investment 2019, 235
World Health Organization, 17, 158
World oil consumption per capita, 183*f*
World Ozone Day, 186
World War II, 181, 259
World's population, 9, 21, 31, 75–76,
203–204

Y

Yom Kippur War, 181–182

Printed in the United States
By Bookmasters